T0059840

ROCKS AND ROCK FORMATIONS

ROCKS AND ROCK FORMATIONS

A KEY TO IDENTIFICATION

JÜRG MEYER

Translated by Mark Epstein

Princeton University Press
Princeton and Oxford

Jürg Meyer is a professional geologist and licensed mountain guide, as well as lecturer and author in the areas of geology and the environment. He lives in Switzerland.

Published by Princeton University Press
41 William Street, Princeton, New Jersey 08540
6 Oxford Street, Woodstock, Oxfordshire OX20 1TR
press.princeton.edu

ISBN (pbk.) 9780691199528
ISBN (e-book) 9780691217550
Library of Congress Control Number: 2021932107
British Library Cataloging-in-Publication Data is available
Editorial: Robert Kirk and Abigail Johnson
Production Editorial: Karen Carter and Kathleen Cioffi
Text Design: D & N Publishing, Wiltshire, UK
Diagrams by Siegel Konzeption | D-Stuttgart
Cover Design: Ruthie Rosenstock
Production: Steven Sears
Publicity: Caitlyn Robson and Matthew Taylor
Copyeditor: Laurel Anderton
Cover image by Shutterstock
This book has been composed primarily in various weights of Meta Pro
Printed on acid-free paper. ∞
Printed in Italy
10 9 8 7 6 5 4 3 2 1

CONTENTS

Preface

The author of this work looks upon its publication with a little trepidation. Learning to identify rocks is a risky business, because identifying them with certainty requires years of experience and can stump experts. I will explain why in the first chapter. For the time being, let me just say that the main reason the world of rocks is so complex is that it is devoid of the concept of species, unlike the worlds of animals and plants. There are no straightforward, unequivocal methods for identifying rocks.

A specialist needs experience and intuition as well as in-depth familiarity with rock-forming minerals, strong powers of deduction, and broad geological knowledge in order to identify unknown rocks. It is not uncommon to hold off on any definite pronouncements until a thin section of rock has been examined under the microscope, a step generally not available to the citizen scientist.

After many years of teaching geology courses, I began to wonder about the creation of systematic keys to identify rocks. Searches on the internet and in numerous textbooks showed that no definitive and complete rock identification key existed for the citizen scientist. After

The fascination rocks exert on us can be experienced firsthand at gravel banks, by rivers, or by the ocean.

experimenting with various concepts, I concluded that the best approach for rocks, one very familiar to botanists, was to use targeted questions paired with grids of possible answers, though admittedly complemented with “loops” that would allow a rock to be correctly identified following different paths. This approach resulted in a final product in which different search paths led to very different rock types being placed next to one another in the same key category. This was unavoidable since the path to identification cannot rely on genetic criteria but must be based exclusively on observable properties.

You can find the necessary foundation, presuppositions, structure, and logic of the key in the chapters that precede it. Following the key are additional specifications and diagrams regarding the characteristics and further classification of the three classes of rock. This volume does not, however, intend to be a work in rock systematics, and it therefore makes perfect sense to use it in combination with one of the many good books that deal with that topic. I wish you enjoyment and success in your adventures in rock identification!

OPPOSITE PAGE:
High-grade metamorphic amphibolite, transformed into an igneous rock. Residual pieces of amphibolite with a biotite-rich reaction rim are contained within newly formed granite (leucosome). Aar Massif, Lötschental (Valais, Switzerland).

why rocks are different

There are many reasons a systematic introduction to rocks is challenging.

Simple Stones?

Small children can distinguish different animal species from one another. Children encounter a wide range of plant and animal species in various ways: in illustrated books, in stories, in school, while drawing or engaged in handicrafts, and so on. From childhood on we are trained to distinguish plant and animal species from each other.

But what is the situation with rocks and minerals? Rocks are simply "stones" or "rocks," and "crystals" are understood to be simply well-formed and transparent rock crystals (quartz). We are not exposed to ways to engage with rocks to identify and classify them. That is certainly one initial disadvantage, even though children are usually fascinated by rocks. Place a small child on a gravel bank, and you will enjoy some quiet for quite a while.

Geology is either not taught at all in schools, or taught very little. Since the subject matter is so complex, students (and frequently teachers) struggle to master the concepts. Even so, we are ready to face the world of rocks.

Flowers, clouds, human beings—everything is differentiated. The rock face: simply gray. Drawing by a ten-year-old child.

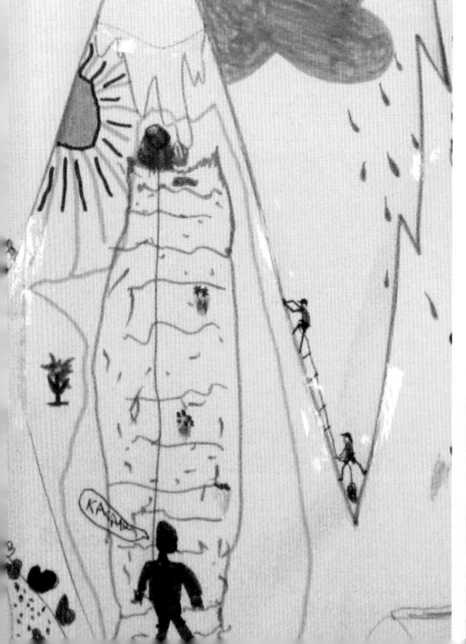

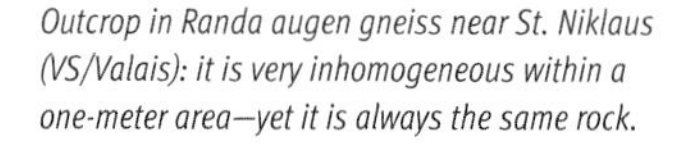

Outcrop in Randa augen gneiss near St. Niklaus (VS/Valais): it is very inhomogeneous within a one-meter area—yet it is always the same rock.

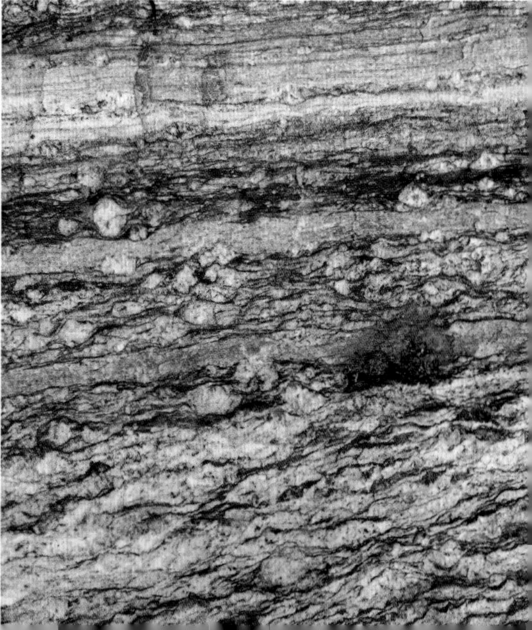

Unusual Space and Time Dimensions

Normally, rocks are not created where we find them today. Instead they have long, eventful histories. If we want to observe and understand rocks, we must include very unusual spatial (in both a geographic, horizontal sense and a topographic, depth/vertical sense) and temporal (time) dimensions.

Mix It Up at Will

In the case of plants and animals there is a clear distinction between species, from genus all the way up to the level of kingdom. The situation with rocks (and minerals) is completely different. You can find continuous transitions between rocks, as well as virtually any kind of combination or mixture. You can find lime mud deposited in an ocean basin, with sand from a nearby estuary mixed on top of it—and these combinations can give rise to very different rocks and rock formations.

Spatial Variability

Rocks can exhibit changes in their composition in all directions; in the previous example of the ocean basin, you can find pure lime mud deposited at one location, but only several hundred feet away sand may already be part of the picture, and still farther away clay may also be included in the mix. These combinations lead to situations where the more recent limestone layers, formed from the evolution of lime mud into rock, are either marl or lime sandstone miles away. Where do you draw the line to define a new rock? You can find similar changes within a well-defined sedimentary rock, for example, but also in igneous and metamorphic rocks.

Gabbro from the Aiguilles Rouges d'Arolla in the Swiss Valais Alps with various secondary changes (green plagioclase, rust-colored oxidation around metalliferous minerals, diallagization of augite).(VS/Valais): it is very inhomogeneous within a one-meter area—yet it is always the same rock.

Secondary Alterations

In the course of their history rocks can undergo constant change, which can also be due to the influence of groundwater and weathering on their surfaces. Such secondary transformations are often realized via the formation of the tiniest mineral grains within the preexisting rock, or minerals: grains that cannot be recognized as such under the magnifying glass, but are seen as mere (dis)colorations. Processes associated with weathering lead to the formation of a weathering carapace, which can look completely different from fresh rock. Coatings of lichen and algae can also transform a rock's appearance.

Nomenclatural Chaos

To top it all, you can add nomenclatural chaos to all this! There is no nomenclatural system for rocks as there is for plants, for example. The names of rocks were just assigned randomly during the course of research. Moreover, many rocks were given local names, so the same rocks can have different names depending on the region in which they are found.

What Is to Be Done?

Both rocks and geologists seem to have purposely arranged things so as to confuse and mislead. What should be done? Throw in the towel? Place this book in a corner out of frustration?

I naturally encourage you to continue onward and instead take these rock-related difficulties as a challenge, working to decipher the rocks' complex messages step by step in the manner of a "SherRock Holmes," so to speak, and thereby discover the pleasures of geological detective work! With time, you can become skilled in rock identification.

You will soon notice that there are indeed specific characteristics of rocks that you can normally follow. So, for instance, you may continually find the same combinations, changes, and alterations, which, once you have recognized them several times, will no longer lead you astray so easily. Using a combination of clues in space and time, from the very small one that you observe under the magnifying glass to the largest that embraces the entire geological environment, you will usually succeed in pinning down a rock's identification fairly successfully. This key to identification will assist you in this task. The more additional information you have about your rock (location, tectonic setting, etc.), the better. Even so, you will occasionally have to be satisfied with an approximation of the rock family you are encountering. Even the experienced geologist, confronted with a rock in the outdoors, will repeatedly conclude, "I believe it could be this or that. I can exclude these possibilities, but in order to obtain a correct result, I need to take a sample with me to the lab." Any geologist who states the opposite is not really being truthful.

OPPOSITE PAGE:
Banded eclogite facies metabasalt from the high-pressure metamorphic ophiolite sequence from northern Corsica. Blue = glaucophane; green = omphacite; red = garnet.

PART 1
the fundamentals

Rocks: An Overview

Crystal—Mineral—Rock

Everyone knows that minerals and rocks have a lot to do with one another and that minerals and crystals are connected. However, precisely what a "mineral," a "rock," and a "crystal" are is often not so clear. Being able to keep these concepts separate is, however, of the utmost importance when working with this identification key. Here are some concise explanations:

Crystals

Crystals are solid substances with a regular atomic structure and a specific composition of atoms and/or molecules (= crystal lattice). Whether or not this crystal lattice manifests itself outwardly via regular crystalline surfaces is really not important.
Example: ice (frozen water/H_2O). Each piece of ice has a crystalline structure, including the beautifully formed star-shaped snowflake and the round grain of firn snow (granular snow found at the top of a glacier), as well as the irregular chunk of ice from a glacier itself.

Minerals

Minerals are natural, inorganic, solid crystalline bodies. All minerals are therefore also crystals.
Example: quartz (SiO_2). A rock crystal with beautiful, regular surfaces is composed of exactly the same mineral as the irregularly shaped grains of quartz in granite, or the rounded quartz grains of sand, which, after the weathering of that granite and a long journey down a river, accumulate on an ocean beach.

Rocks

Rocks are composed of minerals. Rocks can also contain fossils, liquid inclusions, cavities, glass, residual organic substances—or even pieces of other rocks, such as pebbles in conglomerate.

Solid rock and **unconsolidated rock** are generally distinguished; a bank of gravel by a river is also considered rocks by geologists! Many rocks are made up of such small crystals that they cannot be distinguished as such even under a good magnifying glass: geologists call these microcrystalline rocks. Rocks are divided into three large families: **igneous rocks, sedimentary rocks, and metamorphic rocks**. Rocks are related to each other via geological processes in many diverse ways; a sedimentary rock can become a metamorphic rock, and indeed it can even return to a molten stage, which leads to the formation of an igneous rock.

OPPOSITE PAGE:
ABOVE: *Quartz, here shown as a light brown crystal (smoky quartz) that has been crystallized on top of coarse white quartz (with green epidote).*
CENTER: *Mont Blanc granite (French Alps) with an interleaving of irregular gray-brown quartz grains (i.e., crystals).*
BELOW: *Quartz as rounded grains in sandstone.*

Igneous Rocks: In Pluto's Boiler Room

Igneous rocks are formed through the crystallization of molten magma. Magma develops at greater depths in the Earth's crust or in the upper mantle at depths between 30 km and 200 km, with temperatures above 7000°C. Magma contains dissolved gases and crystallized minerals and may include other rocks with higher melting points. As magma cools in a magma chamber deep in the Earth's crust, the developing crystals have plenty of time to reach dimensions of several millimeters or more, until the entire mass crystallizes out as a medium- to coarse-grained igneous rock. Rocks that have formed in this manner are called intrusive or plutonic. The best-known and globally most widespread example is granite.

When magma flows to the surface in a volcano or breaks through via an explosion, it cools very fast and solidifies so rapidly that no slow crystal growth is possible. The result is fine-grained rock known as **volcanic rock** or extrusive rock. Examples are darker basalt and lighter rhyolite. In cases of extremely rapid cooling, no crystals can form; the result is volcanic glass (obsidian, for example). The most significant production of volcanic rocks occurs, invisibly to us, in the midoceanic ridges, where each year, worldwide, enough basaltic magma is produced to cover 450 American football fields with a mile-thick layer, which is deposited as pillow lava or basalt breccias.

Sedimentary Rocks

Sedimentary rocks are formed from deposits at the Earth's surface, which are subsequently transformed into solid sedimentary rock (in a process called "lithification") as a consequence of further superposition with other deposits. Many sedimentary rocks can be distinguished thanks to their stratification, which normally was originally horizontal. The exceptions are cross beddings in river and desert sandstones. Sedimentary rocks can be roughly classified by considering the conditions of their formation.

Clastic Sedimentary Rocks: result from the mechanical transportation and deposition of the products of weathering and erosion, such as rock particles and mineral grains, by water or wind. **Examples:** sandstone, mudrock (argillite), conglomerate, loess, diamictite (materials from moraines).

Biochemical Sedimentary Rocks: incorporation of materials dissolved by water and other particles into the shells or skeletons of living organisms, as well as their accumulation, rearrangement, and deposit. **Examples:** most limestones; radiolarite.

Chemical Sedimentary Rocks: inorganic precipitation of ions dissolved in (sea)water via supersaturation as salts. **Examples:** rock salt, gypsum, potassium salts; dolomite in part.

Depending on the areas in which the deposits are located, very specific sedimentary rocks and sequences are formed. A first major classification is based on whether the sedimentary rocks were formed on solid ground (terrestrial) or in the ocean (marine). The sum of all the characteristics that make up a sedimentary rock is known as its "facies" (bodies of sediment that are recognizably distinct from adjacent sediments). So we speak of a "fluvial (river) facies" or a "shallow marine facies."

Metamorphic Rocks

Metamorphic rocks develop via the recrystallization of existing solid rocks—that is, without melting processes, but all in solid state—when they are subjected to temperatures or pressures in the Earth's crust that are higher than those that first caused their formation. These conditions are most often reached because of the subsidence of rocks in the Earth's crust following tectonic processes. Temperatures and pressures in the Earth's crust both rise as depth increases, and the increase in temperature (the geothermal gradient) can vary depending on the geological circumstances. The temperature and pressure ranges for rock metamorphism begin around 1500°C/1,000 bars and can reach 8000°C/35,000 bars. The temperature-pressure ranges for rock metamorphism are arranged into fields named after basalt rocks (see p. 181). These fields are called "metamorphic facies" and are not to be confused with the facies concept related to sedimentary rocks. Two main types of rock metamorphism are distinguished thus:

Regional metamorphism: tectonic processes either in subduction zones or in the process of mountain formation cause rocks to be buried deep within the Earth's crust with higher temperatures and pressures. This is by far the most frequent case of rock metamorphism.

Contact metamorphism: rocks are heated by magma intrusions, a process that is not tied to subduction to greater depths.

Regional metamorphism is tied to greater pressures and usually also to deformation processes. These can lead to foliation in cases where the rocks contain platelet-shaped minerals from the family of sheet silicates (mica, chlorite, serpentine, clay minerals, etc.). The crystal platelets orient themselves perpendicularly to the greatest pressure or along a shear surface. The foliated structure must be clearly distinguished from sedimentary stratification since the latter is evidence of a very different geological process, but this is not always obvious in the field.

Demarcation Issues

You might guess from what you have read so far that we are confronted with a classification problem in the world of rocks: How do we separate one rock type from another—since the concept of clearly distinguishable species does not exist, and all kinds of mixtures and transitions are possible?

Geologists solve this by two methods: First they compile graphic diagrams for many rock groups that comprise mixtures of different components (minerals, particles, etc.), and within

these they identify fields and assign names to them. The best-known examples are probably the "Streckeisen Diagrams" for plutonic and volcanic rocks (see pp. 175 and 177). Thanks to these diagrams you can deduce, for instance, that a rock is granite even though it may look significantly different from another example. Second, geologists have also defined the concepts "group," "formation," and "member," which are used especially, but not exclusively, for sedimentary rocks. For our purposes the "formation" concept is especially important. It allows us to define a rock unit that can be well demarcated in the field and well represented as a unit on a map. The majority of known rocks are actually formations that can include significant variation in individual rock types.

OPPOSITE PAGE:
ABOVE: *Igneous rock. A meter-wide outcrop in the pluton of the tertiary Bergell granite (southern Graubünden, Switzerland). Different phases of igneous activity result in different types of granitic rocks.*
CENTER: *Sedimentary rock. Sandstone with cross bedding that might lead you to deduce that it is a deposit in a river. From the Triassic variegated sandstone of the Pfälzer Wald, Germany.*
BOTTOM: *Metamorphic rock. An outcrop approximately 50 cm wide in banded amphibolite from the crystalline basement of the Aar Massif at Susten Pass, Canton Uri, Switzerland. The rock was probably formed from very old volcanic tuff deposits with distinctive alternations of basaltic and rhyolitic tuff. Both metamorphism and simultaneous folding took place in the amphibolite facies.*

Rock-Forming Minerals

The World of Minerals: Building Materials for Rocks

When minerals are the topic of debate, most nonexperts think of beautiful crystals as they appear in alpine fissures or round "vugs," with well-developed exterior surfaces and wonderful colors and shapes. Those beautiful crystals in rock shops were crystallized within a cavity filled with liquid. Most crystals, however, grow in competition with other minerals and, as a consequence, in the presence of many other crystals. This situation normally gives rise to a space problem, so that the growing minerals start to impinge on one another, intercrystallize, and then merge with each other: a rock is formed! Other than their outer appearance, a crystal in a rock shop or museum and a crystal of the same mineral formed within a common rock have exactly the same properties and the same crystalline structure; they simply developed under different circumstances. The roughly 4,000 minerals known today are classified according to chemical criteria and atomic composition. Only a small number are involved in the formation of rocks.

The Magical Tetrahedron of the Mineral World

Since silicon (Si) and oxygen (O) are by far the most common elements in the Earth's crust—they make up almost 75% of it—it logically follows that most minerals contain a lot of silicon and oxygen and consequently belong to the silicate group. The two ions Si^{4+} and O^{2-} bind to form a tetrahedron-shaped structure $[SiO_4]^{-4}$, the basic building block of all silicate minerals. You could almost say that this tetrahedron constitutes the foundation stone of solid earth; it is the "magical tetrahedron" of the mineral world. The $[SiO_4]^{-4}$ tetrahedron is to the world of rocks and minerals what DNA structure is to the world of the living. It possesses four negative charges and can bind itself in many different ways to other $[SiO_4]^{-4}$ tetrahedrons and other elements; so among other forms, it can take the shape of a framework, sheet, chain, ribbon, or island. Depending on the type of bond, silicate minerals are known as framework, sheet, chain, ribbon, or island silicates. In addition to the silicate minerals, there are other groups of minerals—for instance carbonates, oxides, and sulfides—in which important rock-forming minerals can also be found.

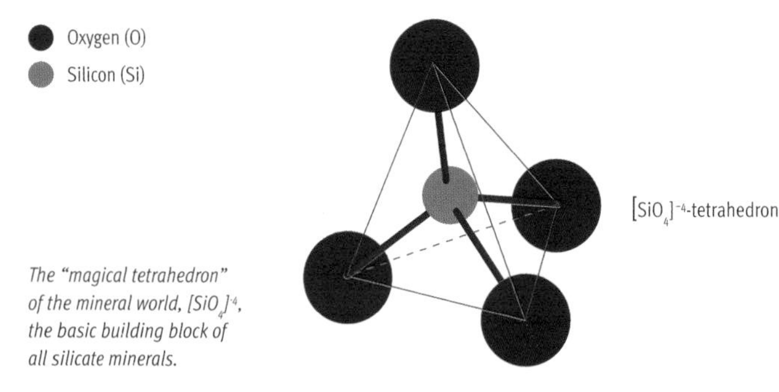

The "magical tetrahedron" of the mineral world, $[SiO_4]^{-4}$, the basic building block of all silicate minerals.

The Great Mixing of the Mineral World

Many minerals can exchange specific building blocks (ions) with other similar ions; this exchange is called "solid solution." Here is an example: the mineral olivine has a pure magnesium variant that goes by the name of forsterite (chemical formula: $Mg_2[SiO_4]$), and a pure iron variant that goes by the name of fayalite ($Fe_2[SiO_4]$). Since Fe and Mg have similar properties, any hybrid forms of these two variants (end members) are also possible, for instance an olivine with the composition $Mg_{1.6}Fe_{0.4}[SiO_4]$. Depending on the chemical composition of the rock in which olivine is formed, it will assume a specific combination composition. In the case of minerals, we therefore speak of a "solid solution series." Most rock-forming silicate minerals exhibit these kinds of "mixing" capabilities. In some cases, for instance the amphiboles and the micas, ions with different properties can be exchanged in combined forms, which leads to a wide variety of end members and solid solutions. This doesn't exactly simplify life for rock enthusiasts. The most renowned solid solution series is that of the feldspars, the most common family of minerals in the Earth's continental crust. In the case of feldspars, specific lattice sites can be occupied by the monovalent ions Na^+ or K^+ or the bivalent ion Ca^{2+}.

Rock-Forming Minerals: The Basics

In order to identify and interpret rocks, it is obviously important to recognize as many kinds of the minerals used to form rocks as possible. The identification key that follows is based on your ability—almost a "core skill"—to distinguish between quartz, calcite, and feldspar, as well as to recognize both of the micaceous minerals muscovite and biotite. For finer levels of identification, it is often necessary to recognize other minerals, such as hornblende and actinolite from the amphibole family, augite and diopside from the pyroxene family, and olivine, garnet, epidote, and chlorite.

Space limitations prevent the inclusion of comprehensive descriptions of the rock-forming minerals. We instead provide a selection in the form of a table and abbreviated presentation, each time referencing examples in the identification key, where they are illustrated with a representative photograph.

An overview of the most important rock-forming minerals

"core skill"
"good to know"
"for advanced learners"

Mineral or mineral family	Group	Chemistry	Most important characteristics
Quartz family Second most common mineral in the Earth's continental crust			
Quartz	Tectosilicate	SiO_2	Colorless to a light smoky brown, conchoidal fracture with a greasy luster; often also as milky white filling in veins and fissures.
Chalcedony	Tectosilicate	SiO_2	Microcrystalline, finest fibrous textural feature of quartz; dense splinter(s)/-ing, often colored (gray-black-rust).
Family of feldspars Most common mineral group in the Earth's continental crust			
Potassium feldspar Orthoclase	Tectosilicate	$KAl[Si_3O_8]$	Colorless to milky white, often with a secondary reddish-brownish-orange coloration (iron hydroxides); mostly semitransparent to opaque; glassy luster, if colored, usually dull. Good fissility, with glassy luster on the cleavage surfaces.
Plagioclase	Tectosilicate	Solid solution series between Na and Ca	Fresh colorless to milky white; often with a yellowish-greenish secondary coloration (very fine inclusions of sericite, chlorite, epidote, etc. ⇢ saussuritization; see p. 44).
Albite	Tectosilicate	$NaAl[Si_3O_8]$	In fresh condition milky white; good fissility, often recognizable.
Family of the feldspathoids (= foids) The SiO_2-poor relatives of the feldspars			
Nepheline	Tectosilicate	$Na(K)Al[SiO_4]$	Short-columnar crystals with six-sided cross sections; greasy luster on the conchoidal cleavage surfaces; can look very similar to quartz. ⇢ Hardness test!
Leucite	Tectosilicate	$KAl[Si_2O_6]$	Strong tendency toward idiomorphic structural features; since it grows cubically, often exhibits rounded forms, often beautiful idiomorphic deltoid icositetrahedron (leucitahedron).
Sheet silicates Always flaky and soft			
Muscovite white mica	Sheet silicate	K-Al sheet silicate	Light silvery to slightly greenish flakes with glassy or mother-of-pearl luster; in weakly metamorphic schists, also as very finely flaky coatings with a silky luster (= sericite).
Biotite dark mica	Sheet silicate	K-Fe-Mg sheet silicate	Dark brown to black, flakes with a glassy luster; dark coloration is attenuated as iron content lessens; a strong tendency to secondary chloritization (see p. 43).
Chlorite	Sheet silicate	Fe-Mg-Al sheet silicate	Dull to dark apple green, opaque, with at most a dull luster; small lamellae and very fine-grained dull green fissure coatings, but also as stacks of flakes and in accumulations or even sand-like aggregates.

Hardness	Distribution/occurrence	Examples
7	In many kinds of rocks, second most common mineral of the continental crust.	18a, p. 102 18e, p. 106 26b, p. 130
7	In sedimentary concretions (chert/flint nodules) and as filling for cavities in volcanics.	13g, p. 91 13c, p. 89
6	In many kinds of rocks, especially in acidic plutonites (granitoids) and in orthogneisses; in granitoids often in the form of phenocrysts resembling Lego pieces.	25c, p. 129 26a, p. 130 26e, p. 131 4e, p. 70
6	In many kinds of rocks, above all in basic plutonites, volcanics (the principal mineral in basalts), and amphibolite.	25c, p. 129 27d, p. 133 4d, p. 69
6	Main component of moderately strong metamorphic greenstones/greenschists; in these instances often as pure white filling in veins and fissures.	18e, p. 107 43b, p. 156
6	Foid minerals are present in igneous, SiO_2-poor/Na-K-rich rocks, especially in volcanics; never occurs with quartz, as otherwise feldspars would form.	24c, p. 127
5.5–6		13c, p. 89
2–3	Above all in metamorphic rocks (schists, gneisses), but also in plutonites, especially in pegmatites. Phengite = light green, Fe-containing variety, in high-pressure rocks. Margarite = brittle mica, in metamarls, harder (H = 4), not bendable.	28a, p. 135 29c, p. 138 30a, p. 142
2.5–3	Common in granitic to dioritic plutonites as well as in many metamorphic rocks. Phlogopite = Mg-rich light brown variety, in marbles and ultramafic rocks.	25b, p. 128 30a, p. 142 18d, p. 105
2–3	Very common in weak to moderate grade metabasites (greenschists), also in weak to moderate grade metapelites. In igneous rocks only secondarily as a transformation product of biotite, amphibole, and pyroxene. In alpine fissures often in larger amounts as fine-grained to sand-like aggregates or as coatings on other minerals.	11d, p. 84 29b, p. 137

Mineral or mineral family	Group	Chemistry	Most important characteristics
Serpentine	Sheet silicate	$Mg_3[Si2O_5](OH)_4$	Mostly as dense masses, from black to blackish green to poisonous light yellow. Frequently greasy lustrous wavy planes of motion due to slickenside. Serpentinites often exhibit orange-brown weathering crusts. Very fine fibrous fissure-serpentine = asbestos.
Talc	Sheet silicate	$Mg_3[Si_2O_{10}](OH)_2$	Very soft (scratchable with a fingernail) whitish to greenish masses, also as flaky aggregates.
Chloritoid	Sheet silicate	Fe-Mg-Mn sheet silicate	Six-sided black flakes similar to biotite, but much harder and brittle.
Glauconite	Sheet silicate	Complex composition	Finely radiated aggregates, quite striking under the magnifying glass due to their strong green color.
Clay minerals	Sheet silicates	Mg-Al-H_2O sheet silicates	Never macroscopic crystals, always submicroscopic; earthlike masses to flaky aggregates; vast family of minerals, best known representatives: kaolinite, montmorillonite, bentonite.
Chain silicates: pyroxene family Grainy to columnar			
Enstatite	Chain silicates	Mg-Fe-orthopyroxene	Dark copper-brown grains, often with a schiller on the cleavage surfaces.
Augite	Chain silicates	Fe-Mg-Ca-pyroxene	Black when fresh, becomes brownish or greenish as a result of secondary changes. Diallage = augite variety present in gabbros and serpentinites with a bronze-like schiller.
Diopside	Chain silicates	Ca-Mg$[Si_2O_6]$	Dull to olive-green pyroxene; forms squat prisms with a ± quadratic to rectangular cross section, also often with xenomorphic grains.
Omphacite	Chain silicates	Na-Al-pyroxene	Dull to full green, grainy to prismatic; can be confused with diopside.
Band silicates: amphibole family The world of blades and rays			
Hornblende	Band silicates	Ca-Fe-Mg-Al-amphibole	Black to blackish green, prisms to blades with a glassy shimmer, even large crystals are pretty common.
Actinolite	Band silicate	Ca-Fe-Mg-amphibole	Mostly long blades of full, light to moderately green coloration, often also radial to acicular aggregates.
Tremolite	Band silicate	Ca-Mg-amphibole	Since it is iron free, it is pure white; otherwise like actinolite.
Glaucophane	Band silicate	Na-Ca-Al-amphibole	Dark blackish purple to lavender blue. In very dark varieties the blue-violet coloration is visible only with good illumination.

Hardness	Distribution/occurrence	Examples
2.5–4	Secondary alteration mineral due to transformation from olivine (and partially, pyroxene) at lower temperatures and with water influx ⇢ serpentinization of oceanic peridotite at midoceanic ridges ⇢ principal mineral of serpentinite rocks.	11e, p. 84 12d, p. 86 19f, p. 111
1	In association with serpentinites or their reaction zones with the surrounding rock, where serpentine-talc-chlorite rocks are formed = soapstone/steatite; also in high-pressure metamorphic metabasites.	9e, p. 81 16d, p. 95 23j, p. 126
6.5	Very common in medium-grade metapelites. Mg-rich chloritoids in high-pressure metamorphic metabasites.	29c, p. 138
2	Formed only diagenetically in ocean sediments; especially in sandstones (⇢ greensandstones), marls, and limestones.	18a, p. 103 21b, p. 113
1	Principal component of mudrocks and marls; as the metamorphism increases, a slow transformation into white micas. Also possible as a result of hydrothermal or surface weathering transformations of feldspars.	9c, p. 79
5–6	In ultrabasic plutonites (harzburgite), in high-grade metamorphic granulite-facies rocks, also in metapelites.	12d, p. 87 19f, p. 111
5–6	Present in basic plutonites, principal component in gabbro; and in the volcanics, principal component of basalt, where it often appears in beautiful idiomorphic crystals.	13a, p. 88 19d, p. 109 27d, p. 133
5–6	In high-grade metamorphic impure dolomite marbles and in ultrabasic plutonites (peridotite); the intensely light green variety containing chromium is an important indicator of the presence of diamonds.	18g, p. 107 23b, p. 119
5–6	In high-pressure metamorphic metabasites, where it is the main component of eclogite together with garnet. Varieties that contain chromium can be a light frog green (variety "smaragdite").	23e, p. 122 30h, p. 145
5–6	In granitic and dioritic-tonalitic plutonites as well as andesitic volcanics (often idiomorphic [euhedral] phenocrysts). It is the most important component in amphibolites (metabasalts), with plagioclase. In marly metapelites sheaflike crystals may also be present.	18g, p. 108 19c, p. 109 22e, p. 116 26d, p. 131
5.5–6	In different medium-grade metamorphites, main component of the greenschists; often also found as a fissure filling in parallel-, radial-, or crazily bladed aggregates.	23d, p. 121 30g, p. 145 43g, p. 158
6	In dolomite marbles and higher-grade metamorphic serpentinites.	
6.5	Present in high-pressure metamorphic metabasites. Gives the blueschist facies its name.	29f, p. 141 30h, p. 145

Mineral or mineral family	Group	Chemistry	Most important characteristics
Sorosilicates and ring silicates Important supporting actors			
Epidote	Sorosilicate	Ca-Fe-Al-silicate	Easy to recognize due to its characteristic yellowish green ("pistachio-green") color; mostly in grainy aggregates, but also easily forms longer blades
Tourmaline	Ring silicate	Na-silicate containing boron	Usual black tourmaline (= schorl); black, three-sided prisms and spicules (needles) grooved lengthwise.
Cordierite	Ring silicate	$(Mg, Fe)_2[Al_4Si_5O_{18}]$	When fresh a deep blue, with a greasy luster; as with plagioclase easily undergoes secondary transformations (⟶ greenish-gray "pinite"); often in rounded grainy aggregates.
Nesosilicates Always hard, often also gemstones			
Garnet	Nesosilicate	Fe-Mg-Mn-Ca-nesosilicate	Always round (cubic) crystal forms, often idiomorphic rhombic dodecahedrons; on weathered rock surfaces as "warts." Common garnet = almandine, reddish brown; Mg-rich high-pressure garnet = pyrope, blood red; Ca-rich garnet orange brown.
Olivine	Nesosilicate	$(Mg, Fe)_2[SiO_4]$	Fresh olive green to light grayish green; the more iron present, the more intense the color; often rounded grains similar to quartz, conchoidal fractures with a greasy luster. Susceptible to secondary serpentinization; orange-brown weathering.
Aluminosilicates: andalusite, sillimanite, kyanite	Nesosilicates	Al_2SiO_5	Three polymorphs that are stable at different grades of metamorphism. Andalusite: light pink prisms; kyanite: blue blades; sillimanite: colorless to white; mostly in very fine spicule aggregates (= "fibrolite").
Staurolite	Nesosilicate	(Mg, Fe, Zn)-Al-silicate	Mostly well-developed idiomorphic dark brown prisms and blades; the very common cross-shaped macles are diagnostically important.
Vesuvianite (Idocrase)	Nesosilicate	Ca-rich silicate	Mostly in squat columns with a square cross section, but also in coarse lumps; colors from dark brown to dull green.
Titanite (Sphene)	Nesosilicate	$CaTi[SiO_5]$	Mostly only small crystals, honey brown to brownish green, wedge-shaped crystal forms.
Nonsilicates 1: Elements The most unequal pair of siblings in the world of minerals			
Graphite/ Diamond	Elements	C (carbon)	Graphite: shiny little black scales and scaly aggregates. Diamond: small octahedrons, mostly ± colorless.
Nonsilicates 2: Sulfides Premier class of the ore minerals			
Pyrite Fool's gold	Sulfide	FeS_2	Brass-yellow color and cubic crystal forms, most commonly cubes; easily weathers to "rust-aggregates" (limonite).

Hardness	Distribution/occurrence	Examples
6–7	In moderately strong to high-pressure metamorphic metabasites; often the principal component of greenschists; also present with feldspar and quartz in hydrothermal veins and lodes; present only secondarily in magmatites, among other forms as the finest transformation product in plagioclase (saussuritization).	23c, p. 121 23h, p. 124 29e, p. 140 43f, p. 158
7–7.5	Noticeable in granite-pegmatites; also in the form of fine spicules in metapelites, orthogneisses, and other metamorphic rocks.	28a, p. 134
7–7.5	In medium- to high-grade metapelites, often also in the leucosomes of metapelitic migmatites and migmatic granites.	23h, p. 124
6.5–7.5	Almandine in medium- to high-grade metapelites and in amphibolites, also in pegmatites, more rarely in granites; pyrope in eclogites and garnet-peridotites. Like quartz, garnets are extremely resistant to weathering and can often be found in sands/sandstones.	23c, p. 120 23e, p. 122 23f, p. 122 23h, p. 124 29c, p. 139
6.5–7	Common in basic magmatites (gabbro and basalt); principal mineral in peridotite rocks. Olivine rocks are (re)formed once again from serpentinite in conditions of high-grade metamorphism.	18d, p. 105 19d, p. 110 23b, p. 118 23f, p. 122
A: 6.5–7.5 K: 4.5–5.5 S: 6.5–7.5	Important indicator minerals of metamorphism in metapelites; kyanite also present in eclogite-facies metabasites.	23e, p. 122 29c, p. 139
7–7.5	Quite common in medium- to high-grade metapelites/metamarls.	29c, p. 139
6–7	Only in metamorphic rocks, in medium- to high-grade calc-silicate rocks, in skarns as well as in rocks resulting from the reaction between basalt dikes and serpentinites (⇢ rodingites).	
5–5.5	Common accessory part in intermediate to basic plutonites, but also in amphibolites and other metamorphic rocks.	
Graphite: 1–2 Diamond: 10	Graphite: very common as an accessory part in high-grade metasediments, marbles, among others. Diamond: in ancient kimberlite pipes and rarely as microdiamonds in ultra-high-pressure rocks.	16e, p. 95
6–6.5	Accessory mineral in many rocks; sometimes larger amounts in rhyolites and mudrocks/shales. Important ore mineral in many mineral deposits, not appropriate for the extraction of iron.	17b, p. 97 40e, p. 149

Mineral or mineral family	Group	Chemistry	Most important characteristics
Nonsilicates 3: Oxides and hydroxides Black ore grains			
Magnetite	Oxide	Fe_3O_4	Appears in the form of shiny black grains; the most common and typical crystal form is the octahedron. Magnetite is ferromagnetic and attracts iron/steel objects (including the compass needle).
Hematite	Oxide	Fe_2O_3	Flaky crystals with a steel-gray coloration and a strong metallic luster; very susceptible to surface oxidation/hydration leading to goethite (limonite).
Ilmenite	Oxide	$FeTiO_3$	Black with a brownish tonality and a flat metallic luster.
Chromite	Oxide	$FeCrO_4$	Irregular black grains with a conchoidal fracture.
Limonite	Hydroxide	FeOOH	Mostly a mixture of both FeOOH forms, goethite and lepidocrocite, with clay minerals; earthy, ocher-colored masses.
Nonsilicates 4: Halides Salt of the earth			
Rock salt Halite	Halide	NaCl	In nonarid regions never at the Earth's surface, as it immediately dissolves; present only in deeper layers. Colorless to bluish, mostly in coarsely spathic aggregates, cubic fissility, glassy luster; tastes salty.
Nonsilicates 5: Carbonates In the world of ocean sediments			
Calcite Calcspar	Carbonate	$Ca[CO_3]$	Normally colorless and transparent to semitransparent, but also in all possible colors (often brownish) due to microscopic mineral inclusions. Calcite's most important diagnostic property is its perfect fissility into three planes, obliquely superposed on one another. This is always visible in broken calcites. Calcite reacts strongly with dilute hydrochloric acid.
Dolomite	Carbonate	$CaMg[CO_3]_2$	Looks almost the same as dolomite, somewhat harder; reliable differentiation in the field is possible only with the muriatic acid test (→ basically does not react with a 10% HCl solution).
Ankerite	Carbonate	Ca-Fe-Mg-Mn-carbonate	Depending on iron content, light brownish to intense brown. Iron-rich ankerites easily weather to loose, earthy limonite masses
Nonsilicates 6: Sulfates and phosphates Mineral water			
Gypsum/ anhydrite	Sulfate	Gypsum: $Ca(SO_4)*2H_2O$ Anhydrite: $Ca(SO_4)$	Gypsum: mostly grainy to fibrous, barely scratchable with a fingernail. Anhydrite: grainy, colorless, white, also pink.
Apatite	Phosphate	$Ca_5(PO_4)_3(OH, F, Cl)$	Mostly small, not at all recognizable with a magnifying glass, colorless-prismatic (six-sided prisms).

Hardness	Distribution/occurrence	Examples
5.5–6.5	Most common in serpentinite rocks, mostly as a beautiful small octahedron in a green matrix. Hard and weather resistant like quartz; often as round grains in marine sands.	40a, p. 148
5–6	An accessory mineral in a very large number of metasediments; massed in iron oolites; also formed hydrothermally in veins.	40b, p. 148
5–6	Common accessory ore mineral in darker igneous rocks; together with magnetite also as a heavy mineral in sandstones.	
5.5	Common in ultrabasic rocks (peridotites, serpentinites), frequently in gravitationally formed layers.	
Soft	Either a hydrothermal or a weathering/alteration product of iron-rich minerals/products; only a little limonite is necessary to make a rock look completely "rusty."	9g, p. 82 Photo p. 47
2–2.5	In evaporitic sedimentary rock sequences, often together with gypsum and dolomite, also with other salts (potassium salts); important as a raw material and as a tectonic decollement horizon (→ for example Jura Mountains).	16c, p. 94
3	By far the most important nonsilicate mineral; dominant in marine sedimentary rocks (shells and skeletons of marine animals). Also the principal component of calcite marble and calc-silicate rocks; partly also in deep- to medium-grade metabasites and serpentinites (ophicalcite). Common also as fissure and vein filling in limestones or Ca-rich metamorphites.	17a, p. 96 17b, p. 96 17g, p. 100 29d, p. 140 43c, p. 157
3.5–4	Is formed by diagenesis from tropical shallow-water lime sludges and Mg-rich interstitial waters, which then solidify into dolomite rocks.	12a, p. 85 17b, p. 97
3.5–4	In metamorphic rocks and in hydrothermal veins.	28b, p. 135
Gypsum: 2 Anhydrite: 3–3.5	Quantitatively the most important mineral in evaporitic rocks; in case of burial gypsum becomes anhydrite following dehydration, and close to the Earth's surface the reverse reaction can occur and gypsum can be formed.	9b, p. 78 16a, p. 94
5	In a great many rocks as an accessory part; in sedimentary phosphorite nodules; principal mineral in our bones.	

Rock Textures and Structures

When we are on the lookout for "beautiful rocks," then in addition to colors, we look mostly for textural elements: here a rock with beautiful corrugations, there one with colorful stratification, another with crisscrossing white fissures, or another with wild and colorful structures. The spatial configuration of rocks and their components, from outcrops to hand specimens, is incredibly important for their description, identification, and the interpretation of their history. This is the reason we must be able to recognize, describe, and interpret their most important textural elements. The following table is a summary of the most important concepts relating to texture, which are also useful for describing rocks in the field. These are also used in the identification key.

Highly deformed mylonitic knobby limestone of Cretaceous age at the foot of the Morcles Nappe (Canton Valais, Switzerland), with folds, shear zones, and cross-cutting (therefore younger) calcite fissures.

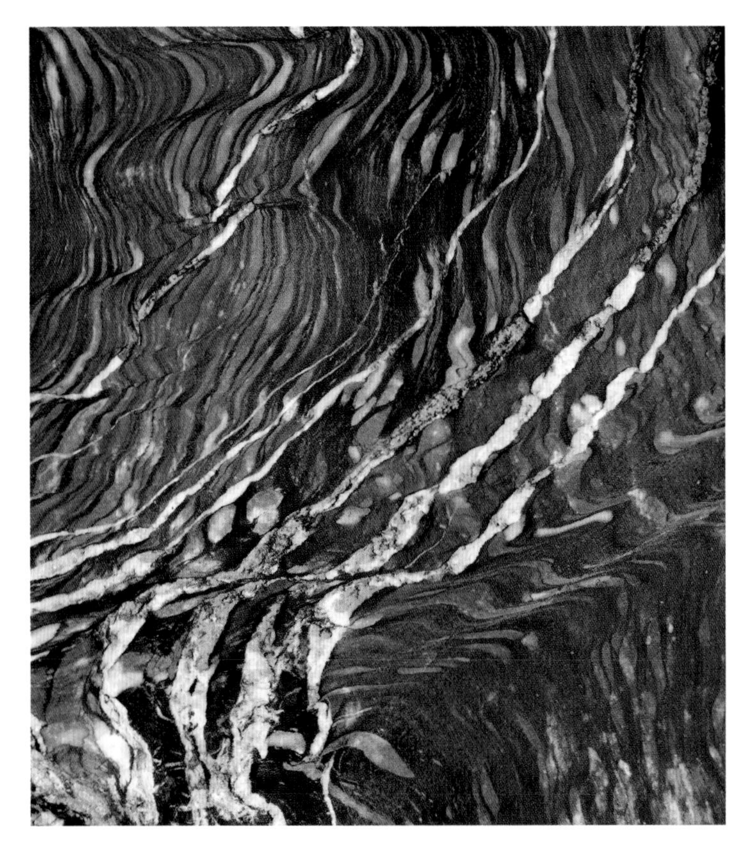

A Spatial Arrangement in the Rock Mass (Outcrop Dimension)

Element	Specifications	Comments
Stratification	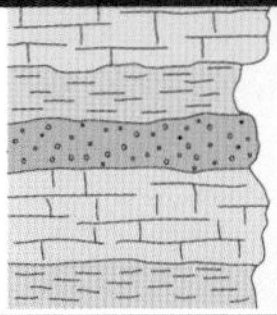	Stratifications are formed as a consequence of changes in the sedimentary deposition.
	Rhythmic bedding	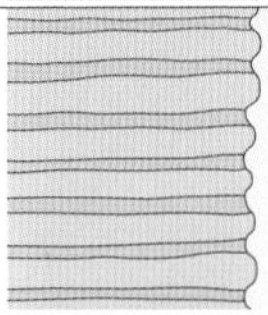 The same type of stratification sequence is repeated multiple times.
	Cross or false bedding	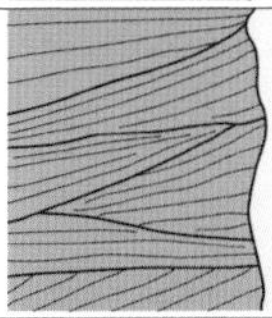 Inclined layers (known as cross strata) form mainly in sandstones that were deposited in rivers (fluviatile) or in desert dunes.
	Graded bedding	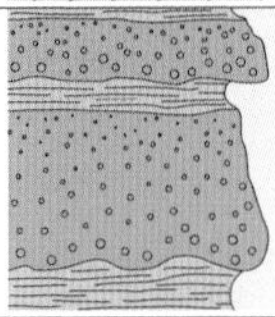 Within a bed the grain size changes continuously from bottom to top.
	Varvity	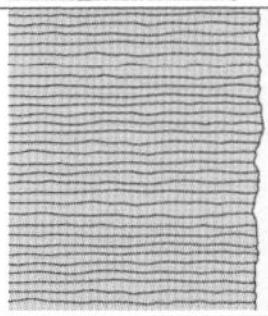 Very fine stratification in the mm range, which exhibits seasonal cycles.
Bedding	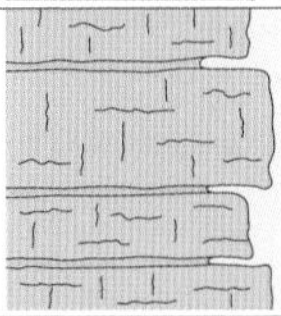	The alternation of variously composed and colored layers. As opposed to stratification, it not only arises from sedimentary processes but also occurs in igneous igneous and metamorphic rocks.
Schistosity		Schistosity differs from stratification in that it is not a depositional structure but instead arises as a result of metamorphic deformation and recrystallization; it is in other words a deformational structure. Schistosity is produced as a result of sheet silicates aligned in a plane, mostly micaceous minerals (muscovite and biotite, among others).
Banding	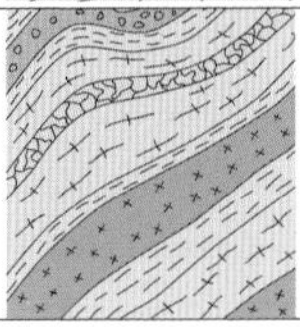	The alternation of variously composed and colored layers. As opposed to stratification, it not only arises from sedimentary processes but also occurs in igneous and metamorphic rocks.

A Spatial Arrangement in the Rock Mass (Outcrop Dimension)

Element	Specifications	Comments
Jointing	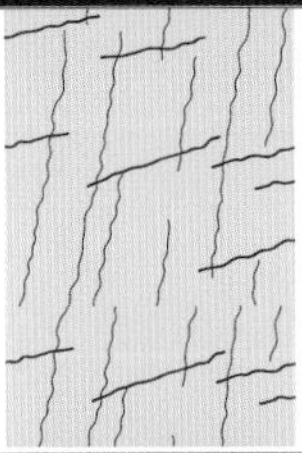	There are almost no rocks that do not show effects of brittle deformation as a result of uplift, burial, or displacement processes. The resulting structures are called "joints." In contrast to veins, they are not filled with minerals. Joints are a consequence of a specific state of tension and for this reason run fairly parallel to one another. Joints are often important sites of permeability for groundwater, as well as sites for the onset of weathering and erosion.
Veins/veining	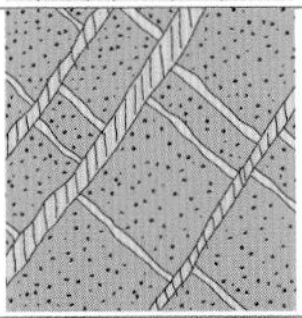	If joints open up and minerals crystallize in these cracks, veins are produced
Shear zones	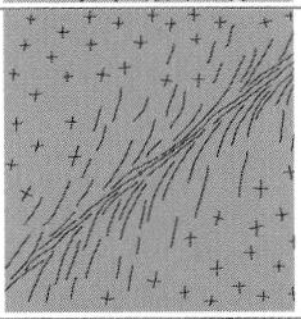	Rocks that are significantly deformed in discrete two-dimensional zones as a result of shear deformations, for instance thrust or normal faults. Brittle and ductile shear zones can be distinguished.
Folding	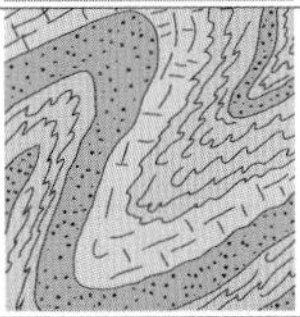	Ductile bending in rocks of all orders of magnitude and intensities.
Boudinage	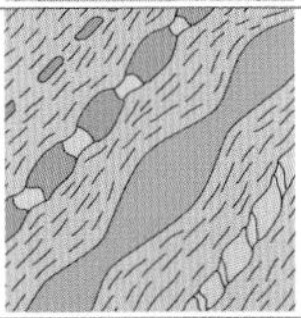	Stretching out of more competent layers between more incompetent ones; in a two-dimensional cross section this leads to sausage-like structures.
Dikes (igneous)	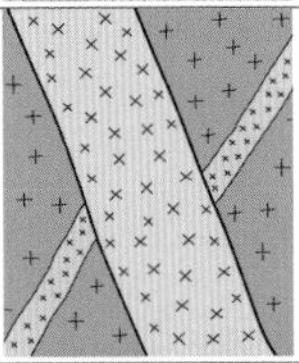	Dikes are a form of magmatic veins: magma penetrates a brittle joint and forms a plate-shaped structure. Dikes can be of formidable dimensions. The most common types are aplites and pegmatites (22f, 28a) in granitoid rocks, and basaltic dikes in basic rocks, as well as rhyolites (4e, 13c, 19a) and lamprophyre dikes (13b).
Migmatite structures	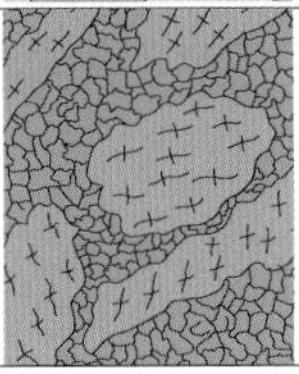	Rocks that heat up to more than 650–7000 C during metamorphism can partially melt. These are referred to as migmatites; they are manifested in distinctive structures that, in the outcrop, consist of an adjacent schistose gneiss and directionless-igneous structures, the latter having formed as a consequence of melting.

A **Spatial Arrangement in the Rock Mass (Outcrop Dimension)**

Element	Specifications		Comments
Special structures in sedimentary rocks	Nodule (nodular) structures/ concretions	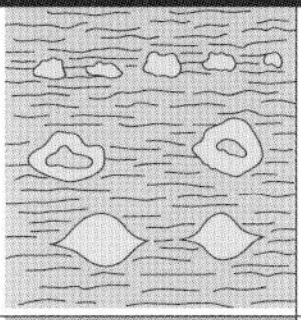	During diagenesis, chemical elements, by diffusion and reprecipitation, can form nodules and concretions of different kinds and sizes in sedimentary rocks. The most common are nodules of the most fine-grained quartz in limestones (silex/chert, 13g).
	Slumps		Fine-grained sediments in a not yet solid condition can form chaotic folded structures on the ocean floor due to (soil) slipping. They can be distinguished from tectonic folds because they are bounded above and below by undisturbed layers of sediment.
	Ripple marks	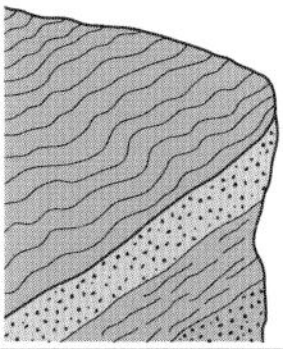	Wavelike structures can result on the upper surfaces of sandstone banks due to the action of water flow(s). These are normally only a couple of cm in height. Depending on how they were formed, current and oscillatory ripple marks can be distinguished. These structures can form in river and flat coastal sands.
	Lumachelles, shelly limestones	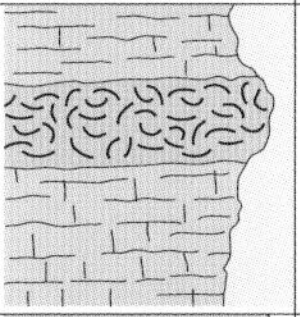	Layers in limestone that consist almost exclusively of the shells or shell fragments of mussels, snails, or brachiopods.
	Tempestites	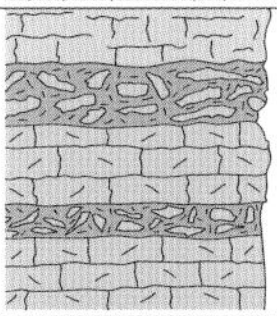	Structures caused by storms or tsunamis in sediments close to coastal areas. In the Alps they are commonly found in dolomites from the Triassic period, where semisolid algal mats were whirled chaotically.
	Flute casts	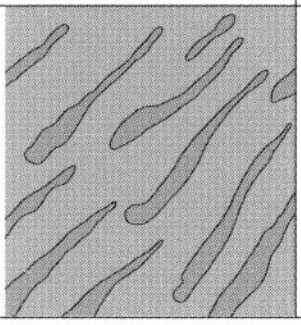	As a result of overflowing turbid currents over fine sediments, flow marks can develop because of local turbulences and remain preserved on the underside of sandstone banks. They show the direction of the flow of the turbid currents. Can be found most often in flysch sequences.

A Spatial Arrangement in the Rock Mass (Outcrop Dimension)

Element	Specifications	Comments
Special structures in igneous rocks	Xenoliths/ "chicken heads"	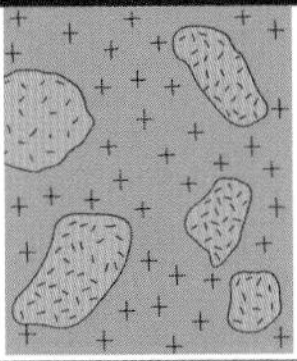 Inclusions of rock fragments in plutonites, for instance of diorite in granite. Orders of magnitude range from cm to 10s of cm. The blocks can be angular or rounded, or even reaction rims with the included rock. If these blocks are harder than the surrounding rock, they weather to the so-called chicken heads loved by climbers.
	Magma mixing	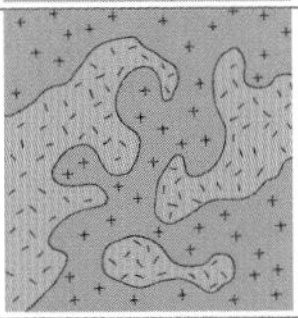 Different magmas can mix in magma chambers without resulting in homogenization and the formation of a new mixture rock. The outcome of this mixing is variable structures somewhat resembling marble cake.
	Flow textures	 Within magma chambers there are magma flows that develop into flow structures. These can be expressed, for instance, in the alignment of potassium feldspar phenocrysts.
	Cumulate structures	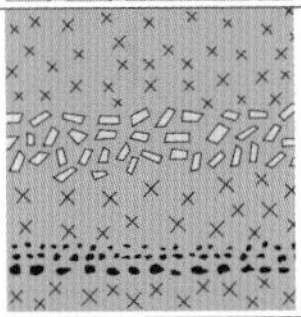 Planar accumulation of specific igneous minerals. These can arise when crystals that have crystallized out of the molten mass then sink because of their higher density.
	Pillow lavas	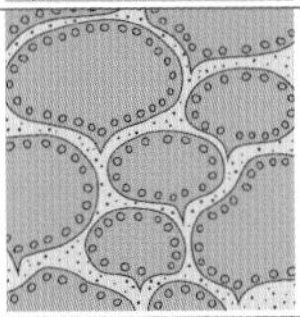 Sausage- to pillow-shaped structures in basalts, with dimensions reaching up to several meters. They are created as a result of the outflow of basalt magma in deep ocean water near midocean ridges and are therefore often encountered in ophiolites.
Special structures in metamorphic rocks	(Contact) aureoles and reaction zones	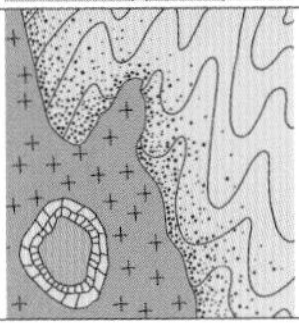 If hot plutons penetrate cool neighboring rocks, they can give rise to contact metamorphic zones (reaction rims) of up to several 10s of meters in size. Since a fluid residual phase occurs after the magma cools, these fluids can penetrate the neighboring rock and give rise to reaction zones there.

B Spatial Arrangement and Characteristics of Components (Hand Specimen)

Element	Specifications	Comments
Space filling(s)	Compact	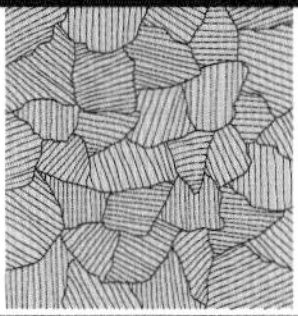 The rock does not contain any cavities or pores.
	Porous	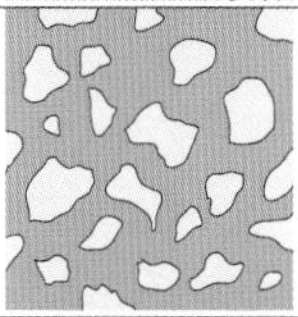 The rock contains cavities or pores.
Grain sizes	Microcrystalline dense ‹ 0.1 mm	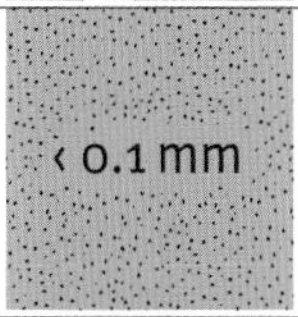 The rock's minerals are so small that even under a magnifying glass they cannot be recognized as individual grains. This is the case with sizes starting at ‹ 0.1 mm and smaller.
	Macrocrystalline fine-grained 0.1–2 mm	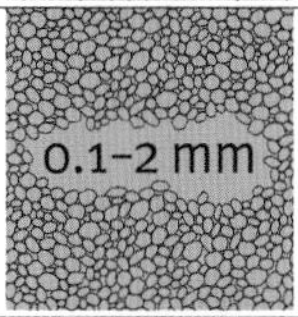 Individual grains can be recognized with a magnifying glass, but detailed characteristics are still too fine to be seen.
	Macrocrystalline medium-grained 2–5 mm	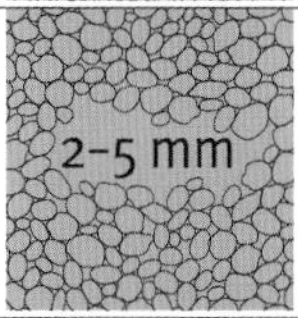 Components can be clearly distinguished with a magnifying glass, and their properties can be clearly seen.
	Macrocrystalline coarse-grained › 5 mm	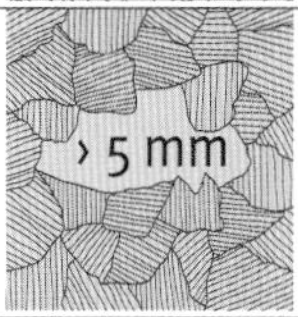 Components can be clearly identified even with the naked eye.
Homogeneity	Homogeneous	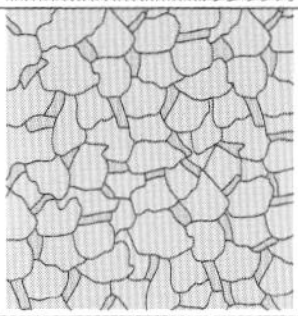 Homogeneous distribution of components, ranging from hand specimens to outcrops.
	Nonhomogeneous	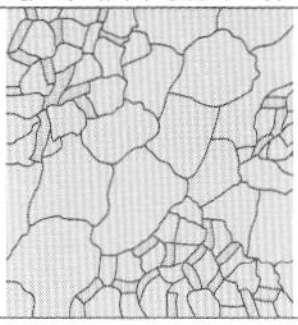 Nonhomogeneous distribution of components, ranging from hand specimens to outcrops.

B Spatial Arrangement and Characteristics of Components (Hand Specimen)

Element	Specifications	Comments
Spatial order	Directionless	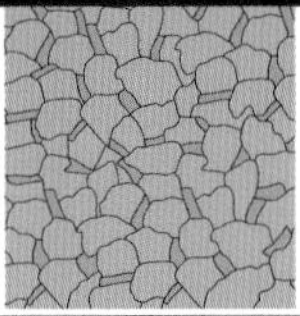 The components are distributed and oriented equally in all directions
	Schistose	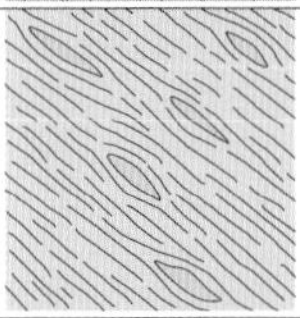 Platelike/foliated minerals are aligned in a specific plane
	Lineated	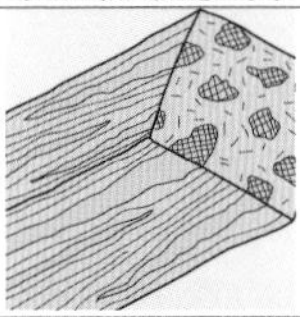 The components are stretched in a single direction lengthwise.
	Stratified	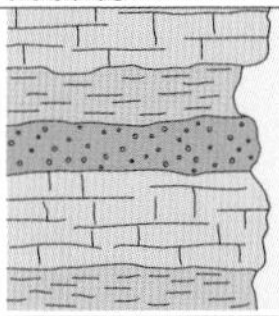 The rock exhibits some stratification.
	Banded	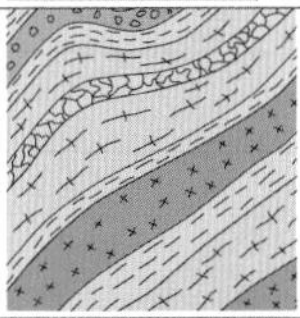 The rock exhibits some banding.
	Folded	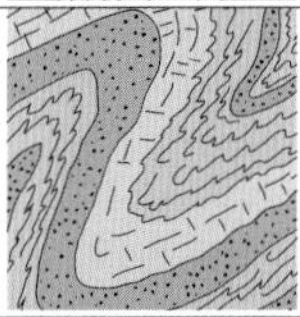 The rock exhibits some folding.
	Crenulated	 The rock exhibits regularly wrinkled and intensive concertina corrugations.

B Spatial Arrangement and Characteristics of Components (Hand Specimen)

Element	Specifications	Comments
Grain size distribution	Equal size	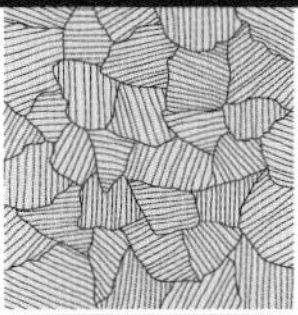 All grains are of roughly equal size.
	Unequal size	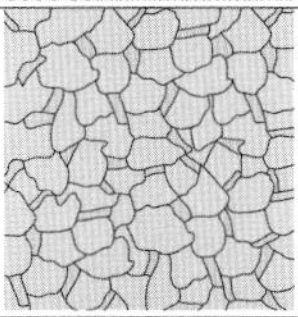 The components, at least in part, are of differing sizes.
Matrix and cement	Porphyritic	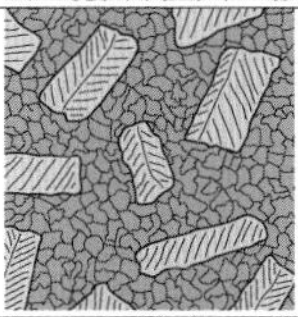 One kind of mineral, usually in very well-formed crystals, is in the form of clearly larger grains in a finer groundmass. Common in plutonic and volcanic rocks.
	Porphyroblastic	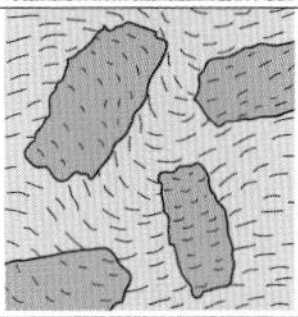 Individual minerals have grown into larger crystals out of the matrix of a metamorphic rock (garnet, for instance).
	Conglomeratic	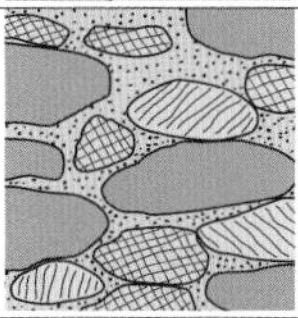 Rounded pebbles lie in a significantly finer matrix (cement). Formed almost exclusively out of river gravel.
	Brecciated	 Angular rock fragments lie in a significantly finer matrix (cement). A result of different processes, most commonly sedimentary-tectonic.

B Spatial Arrangement and Characteristics of Components (Hand Specimen)

Element	Specifications	Comments
Degree of crystallinity	Holocrystalline	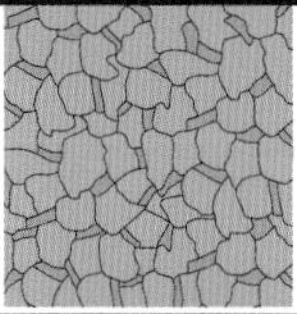 All components of the rock are present in a crystallized form (in the case of very fine-grained microcrystalline rocks, under certain circumstances not very easy to ascertain).
	Semicrystalline	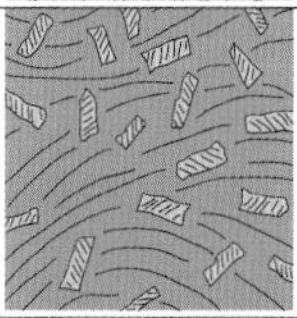 A part of the rock consists of glass.
	Glassy	 The rock consists of glass (usually volcanic).
Textural features of individual grains (minerals)	Xenomorphic	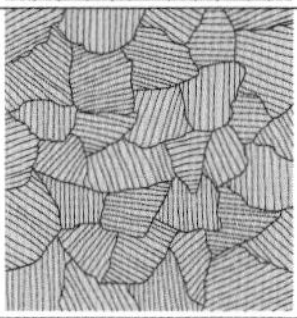 The mineral grains don't in any way exhibit regular exterior surfaces corresponding to their internal crystalline structure but are instead irregularly interlinked with their neighboring grains. This is by far the most common situation.
	Idiomorphic	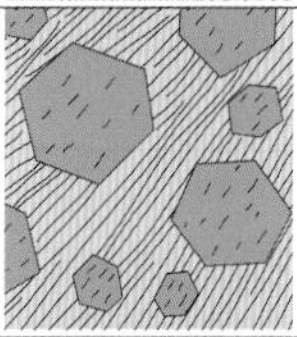 The mineral grains exhibit almost completely regular exterior surfaces corresponding to their internal crystalline structure. Potassium feldspars in granites or garnets in metamorphic rocks are often configured idiomorphically.
	Subidiomorphic	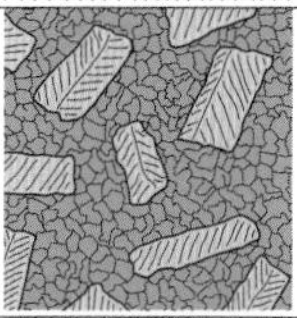 An intermediate condition between idiomorphic and xenomorphic.
	Rounded	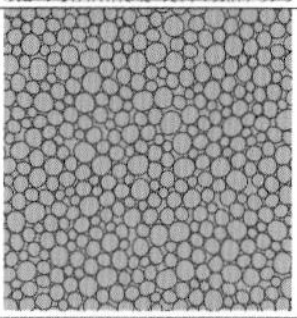 The grains/components are rounded; for instance quartz in sandstone, or gravel in conglomerate.
	Angular	 The grains/components have angular shapes.

B **Spatial Arrangement and Characteristics of Components (Hand Specimen)**

Element	Specifications		Comments
Special textures in sedimentary rocks	Oolites/ooids		Limestone that is built out of small round spheres 0.5–2 mm in diameter. These often allow recognition of a concentric inner structure.
	Algal mats	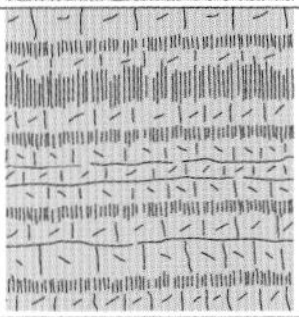	Strata-like layers in the mm range with fine inner structures. Very fine carbonate sludge can be bound to algae along tropical shallow seacoasts. Very common in dolomite rocks.
	Fossils	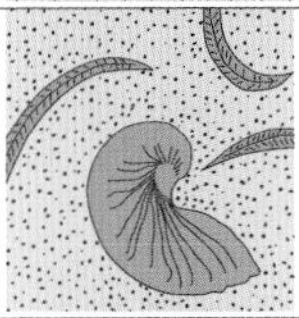	Fossils can form distinctive structures in sedimentary rocks; for instance corals (coralline limestones).
	Trace fossils	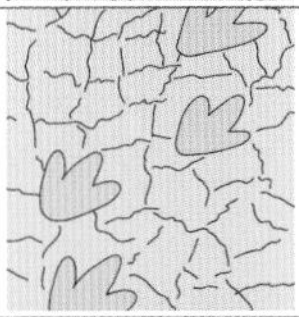	Past life-forms can leave behind not only shells or skeletons but also tracks or signs of their activities.
	Bottom marks	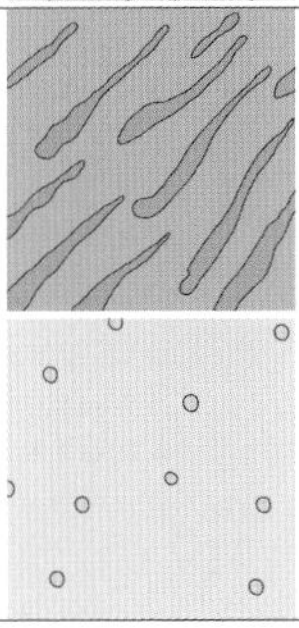	Different kinds of processes can leave traces at strata boundaries; the most important are the flow markings left by flute casts.

B Spatial Arrangement and Characteristics of Components (Hand Specimen)

Element	Specifications		Comments
Special structures in igneous rocks	Spheroidal granites		In some granites spheroidal structures may be present.
	Miarolitic texture(s)	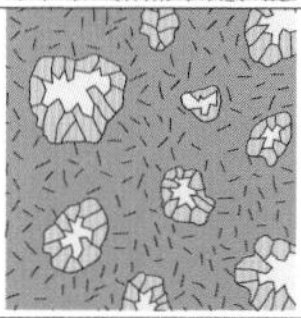	Subvolcanic rocks very often contain small irregular cavities in which some of the minerals can be developed into beautiful crystal surfaces.
Special structures in metamorphic rocks "Blastesis" refers to the crystallization and growth of mineral grains in metamorphic rocks.	Granoblastic	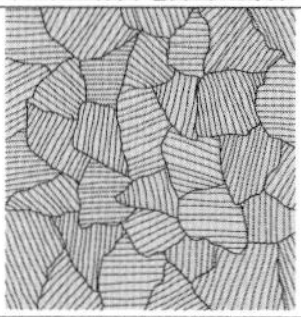	The grains in the structure are about equant, and not aligned; they reflect growth at high temperatures in a balanced structure with straight grain boundaries; common in marbles.
	Porphyroblastic	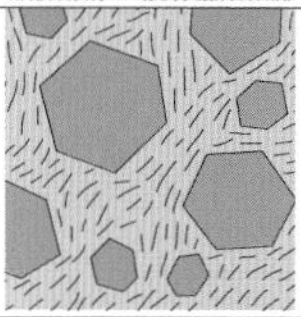	A concept analogous to "porphyritic" in igneous rocks. One kind of mineral grew significantly larger grains than the rest ("matrix").
	Poikiloblastic	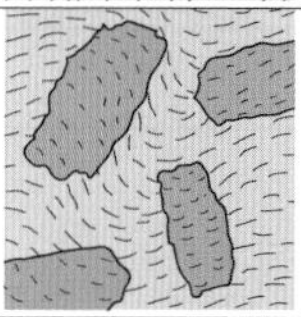	A broadening of the "porphyroblastic" concept: the larger grains overgrow the smaller ones in the matrix; as a consequence, large, often idiomorphically configured crystals riddled with very fine inclusions of matrix minerals are formed. Frequently observed in garnets.
	Reaction rims	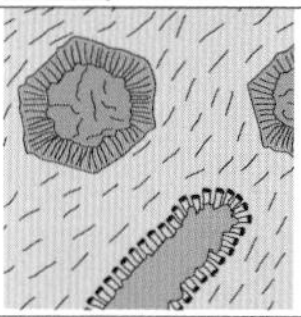	Individual minerals in metamorphic rocks can be subject to reverse reactions during ascent, and therefore at lower temperatures. The extent depends largely on whether fluids penetrated the rock or whether it was deformed once more.
	Pseudomorphoses	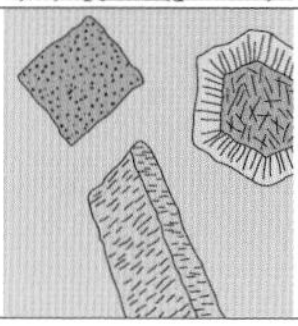	Sometimes during a metamorphic reaction one mineral crystallizes out of another; however, it fills out the appearance and dimensions of the original mineral (in other words the spatial arrangement is the same, but one mineral has replaced the other).

Fossils in Rocks

A whole chain of causal and fortunate circumstances are necessary in order for us to find in a rock the preserved remains or tracks of an animal or plant that lived millions of years ago. The overwhelming majority of dead plant and animal remains decompose, degrade, or are eaten or destroyed. The chances that a living organism will be preserved as a fossil are many million to one! When parts of a living organism are preserved, however, they are usually bones, shells, or teeth. Soft parts are almost never fossilized. For us to be able to admire a beautiful fossil in a rock, the following events need to occur:

— the race between decomposition and covering with sediments has to be won,
— the organism's remains must be decomposed or dissolved in the not yet solidified sediment
— the remains must not be chemically dissolved by interstitial water during the transformation into solid rock (diagenesis),
— they need to survive deformation in the Earth's crust and metamorphic processes,
— they need to be brought to the Earth's surface via uplift or erosion,
— they need to be found, and
— they need to be prepared without incurring any damage.

All Kinds of Fossils

On the one hand there are so-called body fossils: these are the remains of the component parts of animals and plants. These are then further divided according to size into macrofossils (visible to the naked eye) and microfossils (visible only under a microscope). On the other hand there are so-called trace fossils, the great variety of traces that animals have left behind such as tracks, evidence of burrowing, piles of feces, and dwelling structures.

Fossils are, for geologists, bearers of extremely important information. First, they tell us a lot about the conditions in which the surrounding piece of rock was formed. So, for example, corals in limestone indicate that it was formed in the shallow waters of a warm tropical ocean.

The History of Life

The history of life on Earth and the entire evolution of the plant and animal kingdoms can be reconstructed today thanks to fossils that have been collected, cataloged, and classified over the last 300 years or so. The most ancient traces of life are matted structures of bacteria, so-called stromatolites, that can be found even today along the coasts of tropical oceans. They date to almost 4 billion years ago. Since that time, almost the entire history of the Earth has been shaped nearly exclusively by marine microorganisms; it was only about 600 million years ago that larger organisms started to develop, and about 541 million years ago (start of the Cambrian period), life "exploded" into a variety of forms, including larger animals with hard shells, skeletons, and teeth that were suited to fossilization.

Index Fossils as a Chronological Yardstick

Until about 50 years ago the entire classification of the Earth's history was based on fossils; absolute ages in millions of years had not yet been established. Central in establishing this were index fossils. These are groups of animals that were widespread across the globe in many different environments and that developed, in evolutionary terms, as fast and as visibly as possible. So, for instance, the Mesozoic can be subdivided most accurately using ammonites. For modern geologists, microfossils are the most important index fossils.

Fossils as Stone Creators

In some circumstances, fossils can make up the greater portion of some rocks. For instance, the so-called Lumachellen horizons in limestone consist almost exclusively of shell debris from marine organisms such as lamellibranchs, oysters, brachiopods, snails, and so forth. Usually their emergence is a consequence of significant storm events (⇢ tempestite). Then there are layers of limestone, for instance in the Dolomites in Triassic limestone or in rocks from the Jurassic period, which can be understood as essentially fossil reef structures consisting mostly of coral colonies. In condensation horizons where there was no sedimentation for a long time or sedimentation never occurred, large conglomerations of fossils could build up (ammonites, for example). Many sedimentary rocks consist largely of microfossil shells and are always very fine grained. You cannot recognize the shell remains with the naked eye; a microscope is necessary. Among the best known are the white chalk limestones of southern England and northern Germany; these consist almost exclusively of spherical aggregates of coccolithophore shells, which are beautifully structured discoidal shells of marine plankton organisms. These spherical "coccolithospheres" give rise to a somewhat solid, porous rock, which we have used for centuries as chalk for writing.

Secondary Alterations in Rocks

Once a rock is fully formed, it usually has a history of millions of years behind it, up to the point when we hew it off the outcrop. During this history many things can act upon and modify it. Changes taking place beneath the Earth's surface at depths ranging from several hundred meters to many kilometers are called secondary alterations. Rock modifications that occur at or near the Earth's surface are instead referred to as weathering, which will be dealt with in the next chapter.

Most of the rock-forming minerals in plutonic and metamorphic rocks are stable at the high temperatures and pressures of the environment in which they were formed; however, under the conditions that prevail at the Earth's surface they are not—most will degrade to clay minerals, zeolites, and other low-temperature minerals. Granite at the Earth's surface should actually be a piece of clay containing grains of quartz! The reason it isn't depends on energy and kinetic factors. Expressed very simply: the energy threshold for reverse reactions is too high and the reaction speed too slow.

If instead rocks that are formed deep down are affected by groundwater at significant depths and heightened temperatures on their way to the Earth's surface, then these fluids can provoke secondary modification reactions that can transform some or all of the minerals present into lower-temperature minerals. Some hydrothermal back reactions of this kind are particularly widespread and must be recognized, since they can often lead you astray when identifying rocks.

Chloritization

Biotite, since it is a sheet silicate, is particularly vulnerable to corrosion by hot hydrothermal fluids; it is easily transformed into chlorite. This transformation is associated with an increase in green coloration, and "greened" granites are a result of exactly this transformation. But hornblende and augite are also vulnerable to chloritization; you can often find greenish chlorite borders around hornblende crystals in highly metamorphic amphibolites as well as observe chlorite-filled fissures and joints. Even garnets, which are chemically much more stable, can be affected by chloritization at the edges, and in these cases beautiful dark green borders can form around the red garnets. The temperatures required for chloritization lie in the 300–4500°C range.

If a primary mineral is completely replaced by a secondary one in this fashion, a "pseudomorph" is formed, for instance in chlorite pseudomorphs on hornblende.

“Greened” granite from the central Aar region (Swiss Alps). Biotite was transformed into green chlorite, and the plagioclase underwent a whitish-green saussuritization.

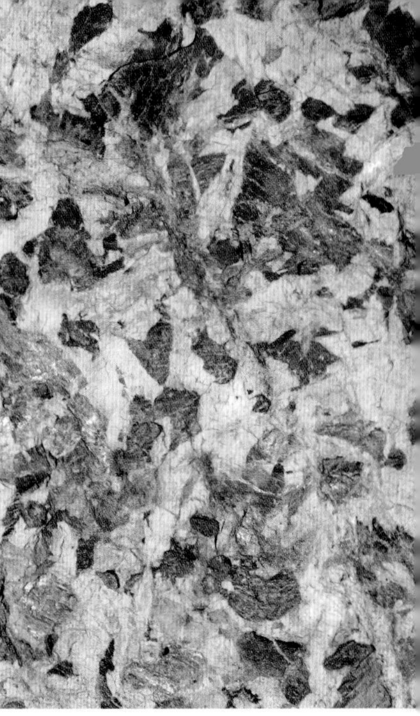

Gabbro from the Aiguilles Rouges d’Arolla (Valais, Swiss Alps) with strongly saussuritized plagioclase, diallagized pyroxenes, and oxidized ore minerals.

Saussuritization

This transformation, named after the renowned alpine geologist and pioneer Horace Bénédict de Saussure, affects plagioclase feldspars; these are also fairly vulnerable to secondary changes by hydrothermal fluids. In this process the grains of plagioclase that appear colorless to white when pristine begin to develop the tiniest new mineral formations at the microscopic level, such as white mica, epidote, chlorite, calcite, and so forth. At the macroscopic level this is expressed as a light yellow-green coloration, which becomes more intense in proportion to the amount of plagioclase that is dissolved. Plutonic rocks with basically completely saussuritized plagioclases are not all that rare. The temperature range is comparable to that for chloritization.

Red Coloration/Oxidation

In some minerals that contain iron, the iron can be dissolved out by the action of hydrothermal fluids and be transformed into iron oxide (hematite) or iron oxyhydroxide (goethite). These processes are associated with a reddish to brownish coloration. The classic example is the large potassium feldspar phenocrysts in granitoids, which are often a pinkish red to a dark rust brown; these are brought forth by the finest hematite lamellae within what is otherwise pristine potassium feldspar.

Alkali feldspar granite with potassium feldspars in vivid red; quartz (gray) and hornblende (black) are pristine; Sinai, Egypt.

Baveno granite (northern Italy) with red potassium feldspar and whitish plagioclase.

Clay Mineral Formation

At temperatures lower than those for chloritization/saussuritization—in other words below 3000°C—many rock-forming silicate minerals are transformed to clay minerals. That is particularly noticeable in the case of feldspars. In some granitic rocks the potassium feldspar degrades into a light white kaolinite aggregate with the consistency of soft earth, which is easy to scratch out with a knife. Plagioclase is instead more likely to degrade to yellowish masses of clay minerals. The area of clay mineral formation interfaces with the surface subjected to weathering, which also leads to the creation of clay minerals.

There are still other kinds of secondary reactions, induced by fluids, within and between rocks; these are usually rare, however, and are described succinctly by the products of their reactions: skarn (23i), rodingite (23i), chlorite talc schist (soapstone, 23j).

Weathering Crusts

The "Makeup" on Rocks

Almost all rocks that we see outside are covered with a more or less thick coating. In geology we refer to this as a weathering surface. This is why, when assessing a rock, we normally need to reach for a hammer in order to observe the pristine ("fresh"), nonweathered rock. Weathering often does not simply stop at the surface; depending on the climate and the type of rock, it can penetrate even meters deep into the rock—granites in the moist tropics, for instance, can exhibit a lateritic weathering surface up to several meters thick. In addition to these chemical forms of weathering, mechanical weathering, for instance from frost shattering, can also be present. Since this second form of weathering does not really affect either a rock's appearance or our approach to a single piece of rock, we will not consider it further. In the case of chemical weathering, the minerals in the rock are transformed by chemical processes. In central Europe's temperate climate three processes predominate: the hydrolysis of silicate minerals, the dissolution of limestone, and the oxidation of ferrous minerals.

Hydrolysis of Silicate Minerals

Both rain and groundwater can impact silicate minerals. When this water is acidic, as a result of its CO_2 or pollutant content, rock degradation is accelerated. The minerals are first dissolved in water, and from these clay minerals are then formed. Feldspars, micaceous minerals, amphiboles, and pyroxenes are the most susceptible to transformation by hydrolysis. Other silicate minerals are chemically very stable and are not at all impacted. Quartz and garnets belong to this group, and they manage to reach the sea as grains of sand unaffected.

Dissolution of Limestone

Rainwater always contains a certain amount of CO_2, which is dissolved as a weak acid in the form of HCO_3 molecules. Limestone (or the rock-forming calcite) is chemically affected by this light acid and goes into solution. This leads first to the formation of fine water channels, which can grow to become very irregular, creviced and corrugated surfaces. These can reach into subterranean systems of hollows and caves, with their respective collapse sites and dolines or sinkholes. Together all these processes are known as "karstification."

Oxidation of Ferrous Minerals

In most ferrous minerals—for instance biotite, amphiboles, and pyrite—iron is present in the divalent "reduced" form Fe^{2+}. As a result of contact with oxygen-rich water, this iron can be dissolved out, oxidized, and precipitated as iron hydroxide, FeO(OH) (goethite, lepidocrocite). "Rusted" rocks of this kind are common: only very few ferrous minerals in the rock are needed for a rust-brown, limonite-rich skin to develop after "treatment" with oxygen-rich water.

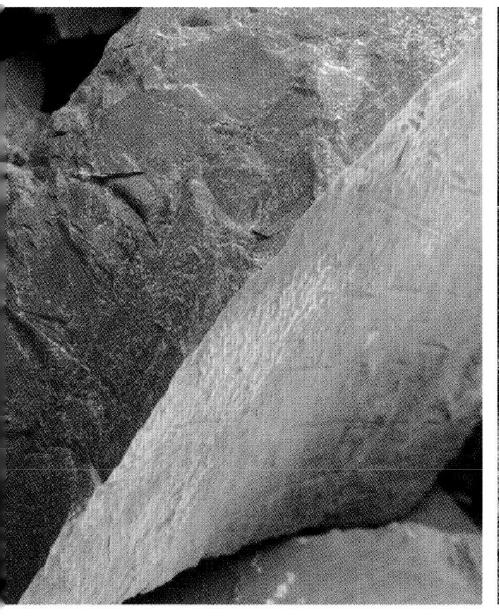

A recent break in gray dolomite rock from them Swiss Alps of Triassic age, showing a marked orange-brown weathering surface.

Geisspfad serpentinite (Binntal, Valais, Swiss-Italian Alps), with green-black color at the surface of the recent break, and orange-yellow weathering edges.

Lichens

In addition to these chemical weathering processes, crustose lichens can change and color rock surfaces to a great extent. The widespread geographic lichen that grows on silicate rocks can color entire areas a greenish hue, and most of the beautiful apparently light to white gray limestone rocks owe this color to extremely thin coatings of crustose lichens.

Crustose lichens also prepare the way for further weathering. With their incredibly fine root threads (rhizines) they can penetrate several millimeters into the rock along the borders of mineral grains, and their organic acids can loosen the association of these grains. This is how the botanical environment grows on rocks, first with coats of algae, often as black "streaks of paint," then with mosses and rock plants, until ultimately the poor geologist is clueless, while the botanist is happy.

Block of gneiss with a "rust skin" among other slightly "rusty" pieces of gneiss and some "rust-free" limestone; Klosters (Canton Grisons, Swiss Alps).

The growth of geographic lichens on granite/gneiss can color large areas of rock green; Rotondo granite, Val Bedretto (Canton Ticino).

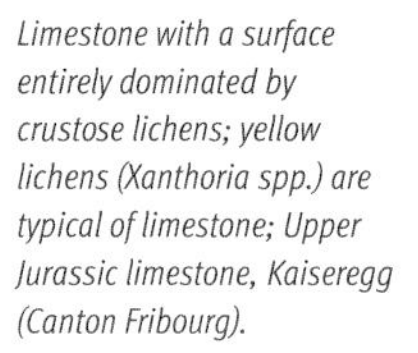

Limestone with a surface entirely dominated by crustose lichens; yellow lichens (Xanthoria spp.) are typical of limestone; Upper Jurassic limestone, Kaiseregg (Canton Fribourg).

Equipment Necessary for Rock Identification

For elementary rock identification you need three important accessories: a hammer, a magnifying glass, and a steel blade.

Without a Magnifying Glass, You're Lost

You will need a handy, compact, folding magnifying glass with at least an 8x–10x magnification and a lanyard. Suitable and economic models can be had for less than twenty dollars.

What the Stick Is to the Hiker, the Hammer Is to the Geologist

This topic can produce hours of discussion and endless leg pulling among geologists! The "softrockers" (sedimentary geologists) prefer a handy geologist's hammer made entirely of steel; the "hardrockers" (geologists of igneous and higher-grade metamorphic rocks) instead prefer a "petrologist's hammer," a long-handled affair weighing about two pounds. Although the author is an avowed "hardrocker" with a petrologist's hammer, he recommends a normal geologist's hammer for the layperson, because this is normally sufficient to chip a fresh corner off any rocks of interest.

The All-Purpose Accessory, the Pocketknife

The abrasive hardness of minerals is a very important characteristic for identification. It can be gauged fairly well using one of the smaller blades on a pocketknife. The knife's blades are also suited to test the finer consistency of minerals (for instance to test whether a specimen really does contain mica lamellae that can be split off).

Several More "Nice to Haves"

For those who would like to pursue their passion at an even more serious level, the following additional items are also recommended:

- diluted muriatic acid (10%) for identifying calcite;
- glass slides for more complete testing of hardness;
- a field book and writing materials to record observations and make drawings or sketches;
- geological maps that, depending on scale, will readily allow you to exclude many kinds of rocks;
- a smartphone, with which you can look up many sorts of geological information on the internet;
- binoculars for recognizing geological structures and rocks in the surrounding area;
- packing materials in case you want to collect rock samples; sheets of newspaper are very well suited to the task.

Approaching Rocks Intelligently

The author has very often noticed, both in courses and on excursions, that someone will pick up a rock and immediately want to know exactly what kind it is. This "instant gratification" approach is understandable, but mostly not productive. Precisely because examining rocks, not to mention identifying them, is such a demanding task, a step-by-step approach is required, followed by a gradual process of acquainting yourself with the item in your hand. The identification key that follows presupposes this approach. A complete and precise observation and description of a rock is an art—and it requires practice! But before we can describe a rock, we must gain access to it.

The Outcrop: Access to the Rock

Geologists constantly talk about "outcrops." They are referring to a piece of bedrock where a specific rock occurs naturally, or "crops out," as they say. This can be in a natural location or in an artificially modified piece, such as a section of road.

The Art of Chipping Off a Good Hand Specimen

As the name suggests, a hand specimen is a piece of rock that easily fits in the hand. This is the approximate size that is usually necessary to describe a rock in all its aspects. Chipping off a good hand specimen for identification purposes is very important—and often not that easy!

It is critical to obtain the freshest possible fracture surface, or surfaces going in different directions. It is all too easy for rocks to develop surface coatings that modify their true character. Various ways to lead you astray, however, also lurk within the rocks themselves: rocks often exhibit very fine microfractures that are not visible from the outside and are covered in very light mineral coatings, which are very easily broken along a fracture by the

The author at an outcrop with pillow lava in the western Alps.

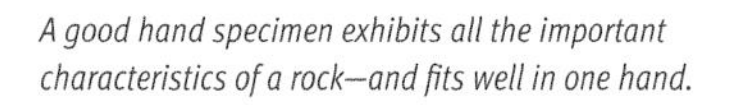

A good hand specimen exhibits all the important characteristics of a rock—and fits well in one hand.

stroke of a hammer—and as a consequence you lose the fresh fracture surface once again! In these cases there is only one solution: to persist, perhaps test different spots on the outcrop, and continue to chip until you have obtained at least two good fracture surfaces. Why is more than one fracture surface necessary? you might ask. When a rock is absolutely homogeneous and does not exhibit any directed structures such as a schistosity or aligned acicular minerals, one surface will suffice. In all other cases, however, at least two surfaces are needed. You may otherwise make classic mistakes when describing the rocks, as in the case of schists and gneisses: platy minerals in these rocks are oriented in a specific direction and concentrated in fine layers. Rocks normally break along these schistosity surfaces after they are struck with a hammer. The result is that you tend to note only micaceous minerals as components of these rocks, almost 100% of the time. If, on the other hand, you take care to create a fracture perpendicular to the schistosity, you will be able to observe that only a small portion of the rock is actually composed of micaceous minerals. A single fracture surface in these cases would therefore be extremely misleading.

Practice, practice, practice! The identification of rocks is otherwise destined for failure from the outset.

Please Don't Fudge: The Correct Use of the Magnifying Glass Is Learned

Just as with chipping off a good hand specimen, correctly using a magnifying glass is something that must be learned and practiced. Those who systematically adhere to these practices will soon see them develop into habits. Here is how to proceed: hold the magnifying glass very close to one eye, close the other, and then move the sample slowly toward the lens until it's in focus. It is important to always have optimal lighting on the surface you are examining. In other words: off with the cap, hold your head high, and turn so either the sun or the strongest source of light at your disposal falls on the sample's surface.

Perfect example of good handling of the magnifying glass.

A Hardness Test for "Hardrockers"

The hardness or, to be more exact, the scratching hardness of minerals is one of the most important characteristics for their identification in the field. In 1822 the Austrian geologist Friedrich Mohs proposed a comparative hardness scale that is both simple and easy to use in the field. This has proved its worth up to the present time, although in the lab it is possible to measure hardness even more exactly using more refined scales (e.g. Vickers hardness test). The practical advantage of the Mohs scale is that you can use everyday materials such as fingernails, copper, steel, and glass as good comparative references in order to roughly classify the hardness of minerals.

Hardness	Mineral	Material	Hardness	Mineral	Material
1	Talc		6	Feldspar	Knife steel 6
2	Gypsum	Fingernail 2, 5	7	Quartz	Window glass 6, 5
3	Calcite/Calcspar		8	Topaz	
4	Fluorite	Copper ca. 4	9	Corundum	
5	Apatite		10	Diamond	

The principle of comparative hardness testing is simple, but its meaningful implementation in the field requires both care and circumspection. The material to be tested must be available with a fresh fracture surface. You must carry out a targeted abrasion, exerting fairly firm pressure with the abrading instrument, and then examine the result with a magnifying glass. Only in this fashion can you determine whether an abrasion was really performed, or whether you are observing only the material abraded from the instrument used.

Beware of the "Sandstone Trap"!

The hardness test is conceived for individual minerals. This is why you cannot simply test the hardness of different rocks in exactly the same manner. You can either recognize the mineral to be examined in the rock in sufficient quantity and in a clearly defined area so that you can proceed with testing there, or, if the rock consists of pretty much only one mineral (for instance limestone or quartzite), you can simply perform the hardness test with the rock itself. At the same time, in such instances you can often walk into a trap that leads to a false result. The most important example is a fine-grained quartz sandstone, where the hard grains of quartz are held together by soft calcite ("cement"). This is a frequent occurrence, for instance with molasse sandstone. If you scratch this sandstone with a steel blade, it will be easily scratched because you are in fact scratching the soft calcite cement, while pushing the hard quartz grains aside. In these cases the opposite test can be helpful: try to abrade a flat steel surface (for instance a knife blade) with the rock—if it contains quartz you will be able to discern traces of abrasion on the blade with a magnifying glass.

Describing a Rock: Something That Has to Be Learned!

Describe something: nothing easier, you think. You couldn't be more mistaken! As someone who teaches courses at a university, the author is always amazed to observe how many difficulties beginners have in describing a rock according to all criteria in a clear and comprehensible fashion. How do you describe the forms of individual minerals, their luster, their fracture behavior, their colors, their structures and textures? Acquiring this capability requires practice—and is decisive for identification. What follows is a simple checklist for the systematic description of rocks.

For some hand specimens you may be able to state something definite about only some of the criteria provided here. Consulting this checklist is worthwhile, however, as part of the preparations for identification: the identification key asks for information from many of the criteria listed here.

Origin of the Sample

Is the sample really from here, or was it transported here?

Kinds of Terrain

How does the rock shape the morphology of the region, the surroundings, and the outcrop?

Characteristics of the Outcrop

Homogeneity of the rock: Is either stratification or schistosity discernible? Are there folds or other structures that resulted from ductile deformation? Are there any brittle deformations such as veins, joints, or faults?

Surfaces, Fracture Behaviors, and Fracture Surfaces

Does the rock exhibit one or more weathering surfaces (patina)? How are they constituted? Does the rock show even the slightest growth of lichen? How does the rock break in the outcrop and in the hand specimen (irregularly, as if made of small clumps, or as small slices)?

Tactile Characteristics

Rocks should be felt and touched! Their surfaces and fracture surfaces often have specific tactile characteristics.

Hardness/Toughness

For hardness testing, consult the previous comments. Toughness says something about how the rock's components hold together.

Specific Weight

You can rapidly tell whether you think a rock is on "the heavy side," or "the light side." You can even compare known rocks of the same size. The "specific weight," or density, is given in grams per cm^3. Limestones and granite, with densities of about 2.8 g/cm^3, are about normal weight, whereas densities above 3 are "heavy" and those below 2.6 are "light." (For comparison, water is 1 g/cm^3.)

Colors and Color Patterns

Is the rock the same color throughout, is it spotted or striped, or does it exhibit other color patterns? How could you describe those colors?

Structures

For a description of the most important structural and textural concepts, see pp. 30–40.

Components

Is the rock microcrystalline? In other words, are its individual components not discernible even with a magnifying glass? If not, how many components can you definitely recognize? Are they minerals and/or clasts? Can you see fossils? Can you estimate the relative proportions in volume % (= modal composition) of those components that are clearly distinguishable?

Rock-Forming Minerals

Try to say or write as much as you can about the individual visible minerals: shape, color, luster, hardness, fissility or breakage; a hydrochloric acid test can also be meaningful. Check the properties against the descriptions in the chapter "Rock-Forming Minerals" (pp. 20–9).

Joints and Veins

When these are present in a hand specimen, they can often help you make inferences about the type of rock in question. The mineral fillings in these joints and veins are often more easily identifiable than the minerals in the rock itself.

Additional Research Methods for Rocks

Given the complexity of minerals and rocks that has been portrayed here, even experienced geologists come up against the limits of the possible when interpreting information in the field—they may take a piece of the rock in question with them to the laboratory. If they are interested in spatial, structural-geological data—such as the orientation of the schistosity surfaces—then they will take a sample oriented with their compass (or smartphone).

An Impressive Arsenal!

Today we have a large arsenal of sophisticated and costly laboratory analyses we can use to examine rocks. We can reproduce their microstructures, especially microfossils, with a scanning electron microscope (SEM) in three dimensions. The chemical composition of rock powder can be determined by means of X-ray fluorescence analysis, and other X-ray methods can be used to determine the crystalline structure of minerals.

FAR LEFT: *Thin section of a micritic limestone with angular grains of quartz and a microfossil of a foraminifer.*
LEFT: *Fluid inclusions of methane from the Lötschberg base tunnel (Swiss Alps) that have formed "negative crystals" in quartz.*

BELOW: *Portable Raman spectrometer for analyzing minerals directly on a rock face.*

An electron microprobe and Raman spectroscopy can be used to examine the exact chemical composition, down to the finest level of detail, of rock-forming minerals. Various radiogenic isotopes in mineral concentrates can be measured thanks to mass spectrometers, thereby allowing determination of their age. Element and isotope distribution has recently been measured in the smallest mineral sections (down to 5 micrometers, μm) thanks to the ion microprobe (SHRIMP), which enables the age of individual grains to be determined. In addition, all possible physical properties such as density, compressive strength, shear strength, macro- and microporosities, conductivity, and so forth can be measured, all of which are important for applied geology. Today some of these devices (RFA and Raman, for instance) are available as handhelds to be used in the field. It's possible that in a number of years this book will no longer be necessary, since the identification of rocks in the field will be accomplished by means of a small device, making everything as simple as calling on a smartphone.

What the Magnifying Glass Was for Sherlock Holmes, the Thin Section Is for the Geologist

There is, however, a comparatively simpler method for researching rocks, one practiced for over 100 years, without which even today's research could not move forward. It is the inspection of the rock with a normal optical microscope (with magnifications of ca. 50x–600x), looking at so-called thin sections of rock that are about 0.03 mm thick. At this thickness, most rock-forming minerals are transparent (the exceptions are the metalliferous minerals such as oxides and sulfides). The thin sections allow you to recognize not only the individual minerals but their textures, internal structures, secondary transformations, deformation structures, displacement reactions, and much more. Thin sections are today, as they were previously, an incredibly important source of observations and interpretations of rock formation.

Over the course of the last decades a separate discipline has evolved from this area that provides additional very important information: the examination of microscopically small fluid inclusions in thicker rock lamellae that are polished on both sides. This in combination with a heating or cooling device, with which the fluid inclusions can be heated or cooled to specific target temperatures, allows conclusions to be drawn as to the kind of fluid being examined, as well as the gases and other substances dissolved in it—very important elements in rock formation.

OPPOSITE PAGE:
A polished piece of "textbook granite" from Baveno (Piedmont, Italy) with its interlocking mineral structures. Gray = quartz, pink = potassium feldspar, white = plagioclase, black = biotite.

PART 2

identification key

Layout and Logic of the Key

This rock identification key is modeled, in terms of both layout and logic, after the well-known botanical identification keys, which consist of mutually exclusive choices and the ensuing branching options. Given this model, however, we are confronted with the difficulties mentioned in the introduction, "Why Rocks Are Different," when comparing rocks to plants:

- there are no firm "species boundaries"; in principle all mixtures and combinations are possible;
- when judged on the basis of texture, similar rocks can appear to be very different from one another (for example limestone can present in forms from microcrystalline to medium-grained textures; colors can also vary immensely on the basis of the tiniest trace admixtures);
- secondary influences and weathering effects can greatly influence the appearance of rocks.

These difficulties were considered in assembling the key, insofar as the same rock can be arrived at following different paths. Moreover, it is possible to backtrack along a path you have already taken, a process facilitated by comments for each entry.

Since identification pathways are based purely on the external characteristics of rocks, the end points of each search pathway do not lead to groupings based on rock systematics; in other words, you may find igneous, sedimentary, and metamorphic rocks next to one another in the key. These are distinguished by means of the corresponding font colors.

The key does claim to at least lead you to the correct rock family, barring cases of very exotic rocks or ones that have been subjected to extreme secondary changes. In some cases the step to the correct specific rock will be easy, but in others you will have to be satisfied with the correct family, since even a geologist would need to undertake more extensive research in order to arrive at an unambiguous identification.

The key was tested by numerous laypersons and scholars and constantly adjusted and improved in the process. It should, in principle, be universally applicable; it does, however, exhibit the influence of a European, more specifically an alpine, geology. Exotic volcanic rocks and rocks characteristic of the old continental shields are sometimes not represented. For instance, the following rocks (in alphabetical order) were omitted: banded ion formation, beachrock, carbonatite, charnockite, fenite, greisen, kimberlite, komatiite, lydite, meteorite, phosphorite, and suevite.

The key is oriented for application with an individual hand specimen. The geological context of the outcrop and the region of origin can obviously be very helpful in narrowing the possibilities, or in supporting decisions pertaining to the choice questions; for some rocks it is indispensable in order to arrive at a conclusive identification.

Prerequisites and Instructions for Use

In order to use the identification key successfully, you must meet the following requirements:

— obtain a proper hand specimen and master the proper use of the magnifying glass;
— recognize the differences between actual rock surfaces and the surfaces of joints, fissures, and veins;
— master the distinctions between quartz, calcite (and where applicable, dolomite), and feldspars; know the various possibilities for distinguishing potassium feldspar and plagioclase; be familiar with the micaceous minerals muscovite (white mica) and biotite (black mica);
— know about other rock-forming minerals; particularly how to recognize amphiboles, pyroxenes, garnets (cf. pp. 22–9);
— understand the most important concepts relating to structure (cf. pp. 30–40);
— have the most fundamental accessories available (cf. p. 49); dilute muriatic acid is not required but is definitely helpful (diagnostically important information relating to muriatic acid is conveyed in the identification key using the symbol HCl + or HCl -);
— exercise caution in applying the hardness test on rocks; always also conduct a scratch test using the rock on the testing material, so as to avoid the "sandstone trap" (cf. pp. 52).

Working with the identification key—exciting detective work.

Patience and Exact Observation!

Working with the key requires patience, especially at the beginning, since all the text relevant to choices and decisions must be read very carefully and compared with observations of the hand specimen. With time, the central choice questions at the beginning of the key will become familiar, and you will be able to proceed fairly rapidly to the later questions, allowing you to arrive at a more concrete identification.

Each choice question is illustrated with one or two photos of relevant rock examples. The decision to use these photos was reached after prolonged hesitation: they can be extremely useful, but you can end up relying on them too heavily, and forgetting that the relevant variation ranges are so significant that a rock could also be exemplified with ten to twenty different accompanying photos. You should therefore always use caution when consulting the photos.

A difficulty that repeatedly turned up in tests of the key was tied to choice question 2, "macro- or microcrystalline." We categorize rocks as "macrocrystalline" even if they exhibit very small grains of less than 1 mm in size; as long as you can still recognize these as grains with a good magnifying glass, you can decide whether they present only one kind of mineral or more. You should therefore always remember that for the purposes of this identification key the concept "macrocrystalline" relates primarily not to size, but to visible mineral grains.

The Colors of the Rock Groups

The rocks are grouped into tables with the following colors:

— unconsolidated material
— sedimentary rocks
— igneous rocks (plutonic and volcanic rocks)
— metamorphic rock
— hydrothermal, tectonic, and special case rocks

Some Pointers on the Nomenclature of Rocks

As already mentioned on p. 12, rock nomenclature is chaotic and is to be understood only in a historical context. You must simply master the most common names over time. A complicating factor is that, depending on the point of view, a rock can be assigned quite different names that are all correct: so, for example, a rock from the uppermost mantle, which consists largely of olivine with some orthopyroxenes, can have the following designations:

— "ultrabasic rock" (a group of igneous rocks with SiO_2 content below 45% overall weight);
— "ultramafic rock" (a group of igneous rocks with dark mafic mineral content of more than 90%);
— "peridotite" (a rock that consists mostly of olivine);

— "harzburgite" (a more precise narrowing down of peridotite for a rock with olivine and orthopyroxenes); and
— olivine-orthopyroxene-cumulate (a rock that was formed via gravity-induced settling of olivine and orthopyroxenes in a magma chamber).

Especially in the case of igneous and metamorphic rocks, names can be made more precise thanks to the use of adjectives or ad hoc composite formulations. This can best be explained with the following two examples.

Example 1: a granite that contains potassium feldspar in the form of large phenocrysts within a medium-grained matrix, as well as a considerable amount of biotite (around 15%) and some hornblende (around 5%) in the dark minerals group, could be referred to as "medium-grained porphyritic, hornblende-bearing biotite-granite"—an expression that is obviously more informative than simply "granite."

Example 2: a highly metamorphic, medium-grained gneiss that has significant folding, and which, in addition to a lot of quartz and some feldspar, also contains in macroscopically visible amounts the following minerals (ca. 20% biotite, ca. 8% hornblende, ca. 5% garnet, and ca. 3% tourmaline), can also be referred to as follows: "medium-grained, folded, tourmaline-, garnet-, and hornblende-bearing biotite-gneiss." In cases like this the accessory constituent designations are placed closer to the mineral's name the greater their proportional share, usually expressed in a percentage.

You can proceed in a similar fashion with sedimentary rocks. You will find out more about nomenclature and further classification of rocks in part III.

But now, get started on your identification adventure! Good luck!

Identification Key with Illustrations

The key allows for the identification of all rocks of a certain significance. Some very special rocks, or rocks that, for example, occur only within the very ancient continental shields, are not considered here (for instance banded ion formation, beachrock, carbonatite, charnockite, fernite, kimberlite, komatiite, lydite, phosphorite, and suevite, among others). Likewise meteorites, except for those belonging to the tectite family, are also not dealt with. These "exotic rocks" are, however, briefly described at the end of the key (pp. 170–1).

"HCl" Column

This column provides information about the rock's reactions to a diluted, 10% solution of muriatic acid (HCl). In the presence of calcite (calcspar/calcium carbonate) the rock reacts with intense foaming. Knowing that in most cases laypersons will not have muriatic acid with them, we still provide this information for those who would like to use this testing agent. These are the possible options:

⊕ clearly reacts with HCl → calcite present in significant quantities
⊖ does not react or reacts very weakly with HCl → no or very little calcite present
⊕/⊖ both reactions are possible, depending on the amount of carbonate in the rock
– the HCl test is not a criterion for choosing among mutually exclusive options

Scale of the Photos

Most of the photos reproduced here are macro shots. For reasons pertaining to both graphic effects and legibility, we intentionally omitted information regarding scale. Most of the photos show a rock surface from about 2 × 2 cm to 5 × 5 cm and therefore the rock appears as you would see it with the naked eye, looking at a good hand specimen.

In exceptional cases the photos also show more extensive images of the rock(s), occasionally even entire outcrops on a scale of meters; this is normally quite clear from the illustration itself. For unconsolidated rocks, entire landscapes are sometimes depicted to show the context of the deposits.

Solid Rock	› 2
Vein fillings (of joints or fissures)	› 43
Unconsolidated rock	› 44

To identify your rock please proceed to chapter/page

Solid Rock

1		Ch./page
a	Normal case. A rock in the hand specimen range that is cohesive and firm.	→ 2/65

Vein fillings (of joints and fissures)

		Ch./page
b	Planar fissure fillings composed of one or more roughly crystallized minerals that traverse the rock. If you do not want to identify the source rock, but rather these fillings themselves, proceed to... **NOT** included in the joint, fissure, or vein fillings is igneous matrix [red font color cf. Original p. 63]. This should be identified by means of the entries under "Solid Rock."	→ 43/156

Unconsolidated rock

		Ch./page
c	Rock in the hand specimen range that is not cohesive and falls apart (rubble, gravel, sand, silt, among others).	→ 44/160

Solid Rock	→ 2
Vein fillings (of joints or fissures)	→ 43
Unconsolidated rock	→ 44

Solid Rock

2	a–f	Ch./page

from 1a

a Grains/minerals are recognizable to the naked eye and with the magnifying glass, even when in certain circumstances they can be very small. You can recognize individual mineral grains and decide whether a hand specimen contains one or more kinds. → **Macrocrystalline rocks** → 14/92

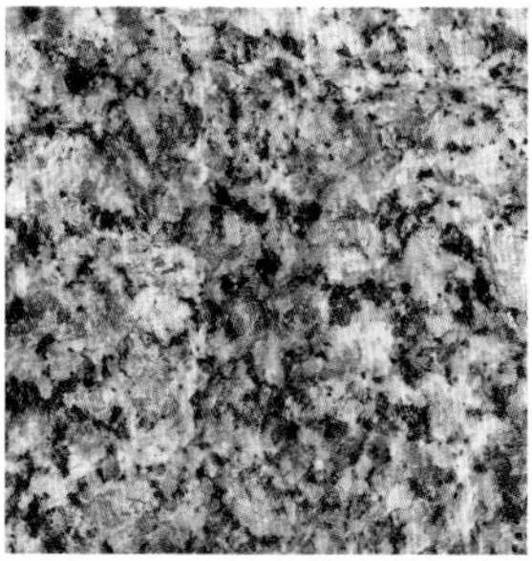

b Grains/minerals are not really recognizable to the naked eye or with a magnifying glass; the rock is very fine grained but not glassy. → **Microcrystalline rocks** → 8/77

c The rock is not schistose and is made up of **components** or **minerals** visible to the naked eye, which are set in a much finer groundmass (cement, matrix, which can be macro- or microcrystalline). → **Composite rocks (components & matrix)** → 3/67

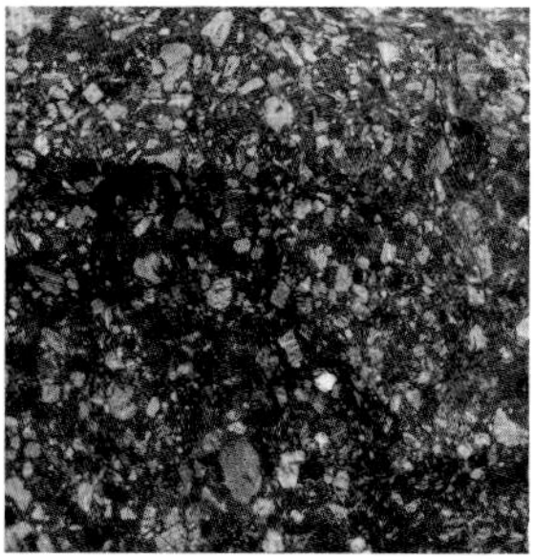

continued overleaf

Ch. / page

d The rock is microcrystalline and from porous to bubbly (it may also contain disseminate crystals or other components). → **Porous and bubbly rocks** → 4 / 68

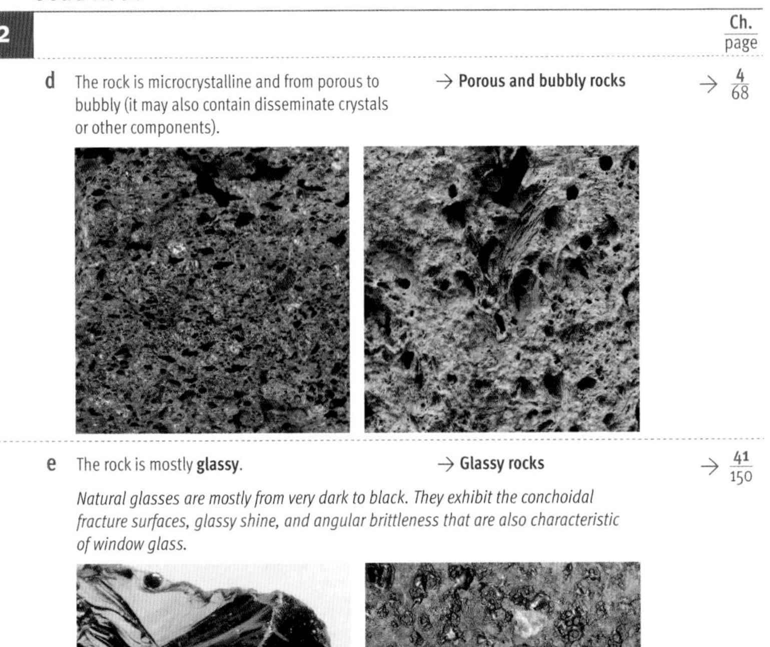

e The rock is mostly **glassy**. → **Glassy rocks** → 41 / 150

Natural glasses are mostly from very dark to black. They exhibit the conchoidal fracture surfaces, glassy shine, and angular brittleness that are also characteristic of window glass.

f The rock is thoroughly veined, and the amount of vein material is significant. The veins are not folded in a ductile fashion; in other words, they exhibit brittle deformation. → **Identify the source rock and the vein fillings separately as distinct rocks** → 42 / 152

Some characteristic kinds of rock do exist, however, which are thoroughly veined and have been assigned their own names. They are listed under 42.

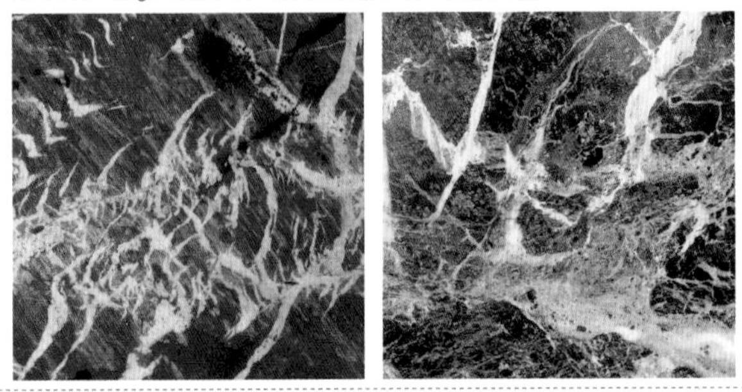

Composite rocks (components & matrix)

3 a–b

Ch./page

from 2C

a The rock is compact, which does not mean that it cannot also contain individual smaller, hollow cavities, but these do not really contribute anything to the true character of the rock

→ 5/71

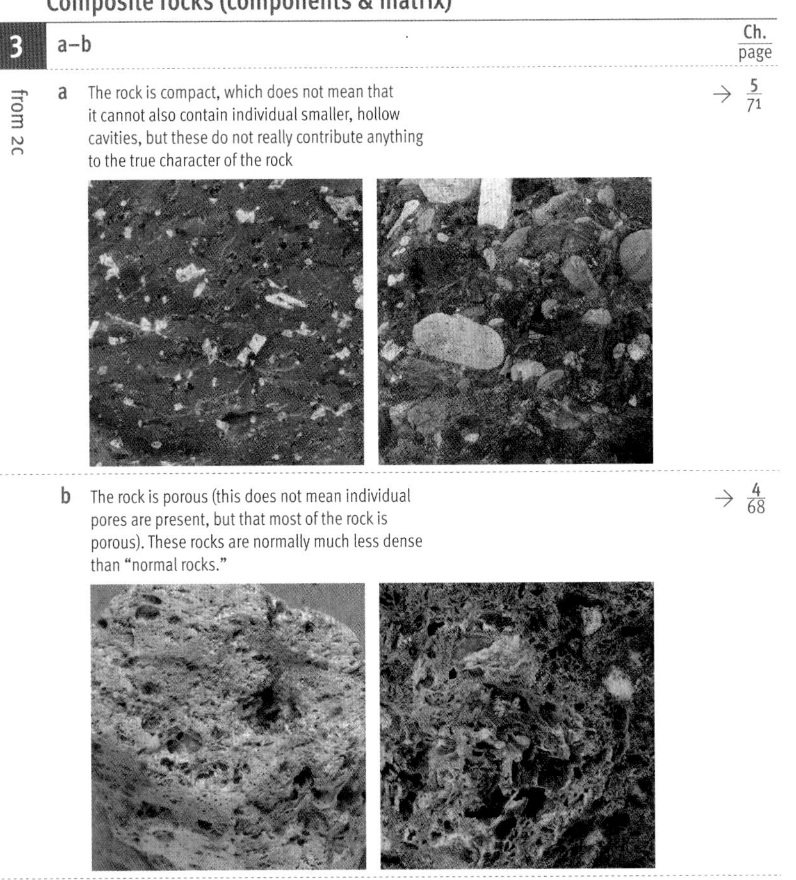

b The rock is porous (this does not mean individual pores are present, but that most of the rock is porous). These rocks are normally much less dense than "normal rocks."

→ 4/68

Microcrystalline—porous/bubbly rocks

4 a–f

from 2d and 3b

a The rock is gray to ocher yellow, from porous to cellular, relatively soft to very soft, often also brecciated (mostly contains pieces of dolomite). → **Cornieule** (Sedimentary rocks or tectonite) ⊕

Cornieules are tectonically influenced evaporite rocks. The holes originate from gypsum, which was dissolved out by groundwater.

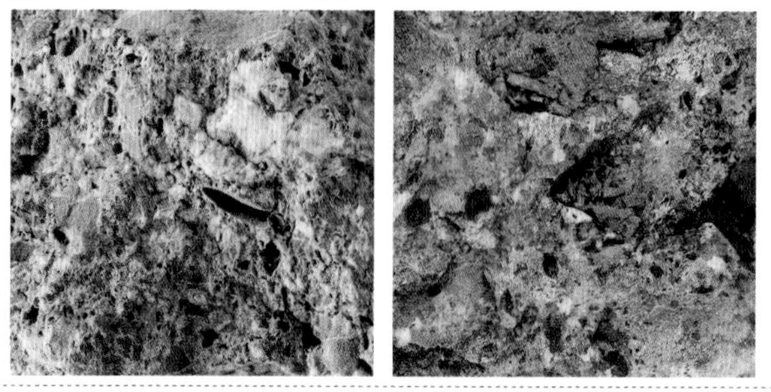

b The rock is gray to ocher yellow, porous to cellular holey, sometimes with encrusted remains of branches or leaves. → **Sedimentary tufa, calcareous tufa,** or **travertine** ⊕

Sedimentary rock. Pay attention to the geological context! Sedimentary tufa was recently formed at calcium-rich springs and waters. In some locations this easily worked stone was used for lintels on windows and doors

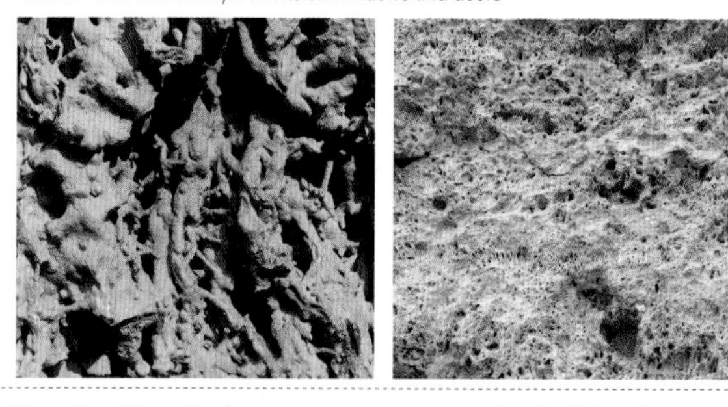

c The rock contains volcanics as components or originates from a volcano complex; the colors can vary (mostly brown/gray tones). → **Volcanic breccia** or ⊕ ⊖

Volcanics, volcano-sedimentary rocks.

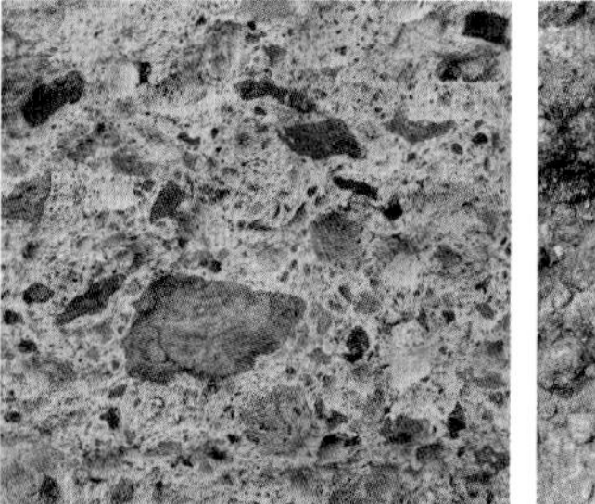

4

c *Volcanics, volcano-sedimentary rocks.*

→ **Hardened lapilli** ⊕
(often in unconsolidated ⊖
rock form)

d The rock is dark gray to black, foamy to porous.

→ **Basaltic or andesitic lava**
(see also 13a, 19c and d)

Differentiation of andesite vs. basalt

→ **Andesite:** possible phenocrysts are plagioclase (transparent when fresh, greenish opaque when transformed) and black hornblende blades. ⊖

→ **Basalt/olivine-basalt:** possible phenocrysts are black augite prisms and/or olive-yellow-green olivine. ⊖

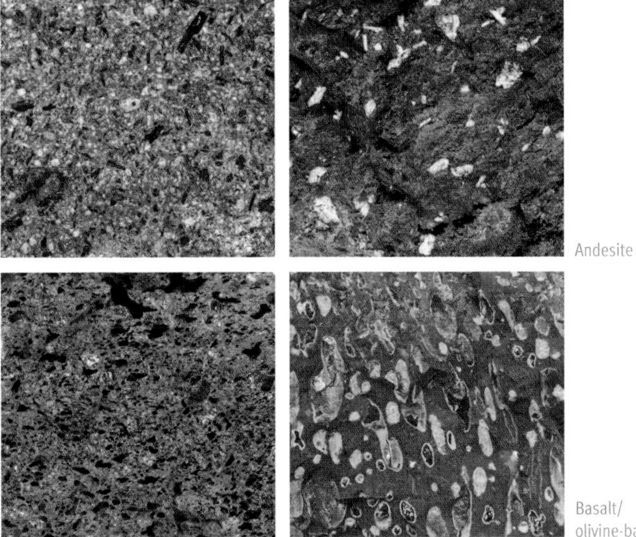

Andesite

Basalt/
olivine-basalt

These kinds of lava can also contain disseminated minerals that were already crystallized in the magma in more or less pronounced idiomorphic shape before the eruption; "amygdales" are also occasionally present (see photo on the right); these were originally gas bubbles that are now filled with minerals.

continued overleaf

Microcrystalline—porous/bubbly rocks

e The rock is foamy and porous; colors are often gray, but rust-oxidation colors are also possible.

→ **Rhyolitic to trachytic lava:** as with (d), disseminated crystals may also be present. (see also 13c, 19a)

→ **Rhyolite:** possible phenocrysts are quartz (round, gray, shiny) and alkali feldspar (white to reddish brown, small blocky crystals). ⊖

→ **Trachyte:** possible phenocrysts are transparent to white alkali feldspars. ⊖

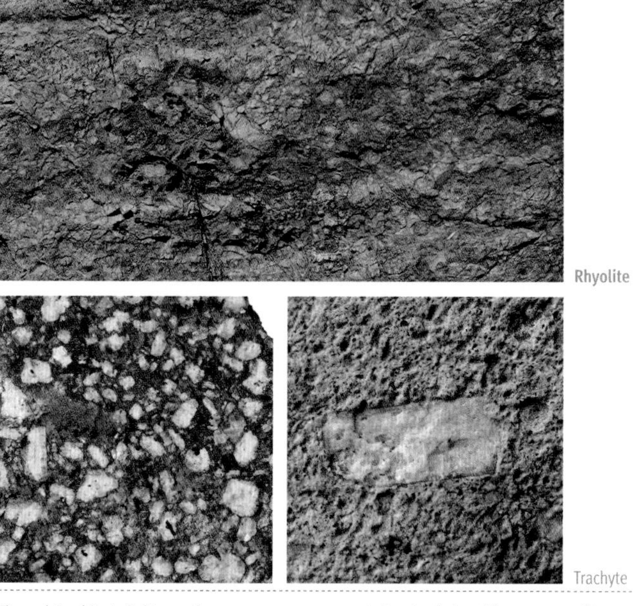

Rhyolite

Trachyte

f The rock is white to light gray, foamy porous, very light (often can float on water), light coloration.

→ **Pumice** (volcanic) ⊖

Pumice is the geological champagne foam: it is explosively expelled, tough rhyolitic lava that foams up and immediately solidifies.

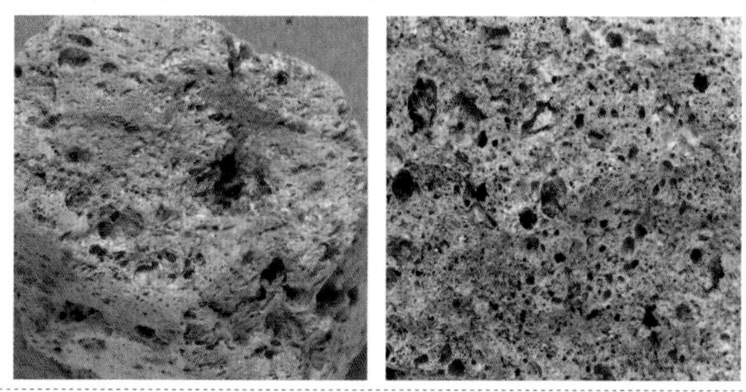

Composite rocks (compact)

5 a–f — Ch./page

from 3a

a Components are round rock fragments (“gravel”). → **Conglomerates** → 6/73

b Components are mostly angular rock fragments (not mineral grains!). → **Breccias** → 7/74

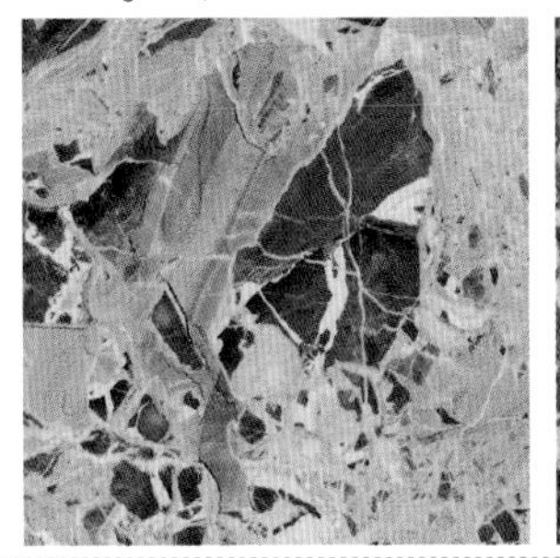

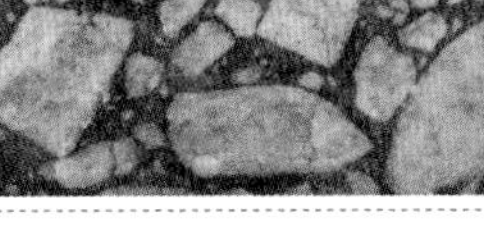

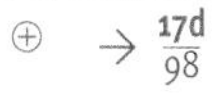

c Rock contains fossils/fossil fragments in a fine-grained, mostly microcrystalline matrix. → **Bioclastic limestone** ⊕ → 17d/98

continued overleaf

Composite rocks (compact)

5

Ch. / page

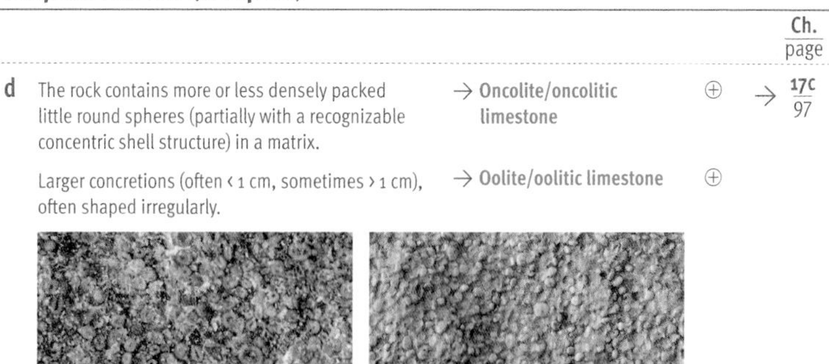

d The rock contains more or less densely packed little round spheres (partially with a recognizable concentric shell structure) in a matrix. → **Oncolite/oncolitic limestone** ⊕ → 17c / 97

Larger concretions (often ‹ 1 cm, sometimes › 1 cm), often shaped irregularly. → **Oolite/oolitic limestone** ⊕

Oolite/oncolitic limestone

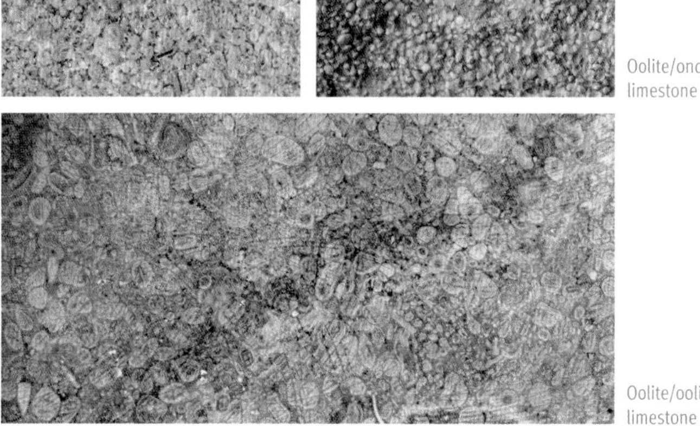

Oolite/oolitic limestone

e Components are individual minerals, often with recognizable crystal shapes, in a very fine-grained or microcrystalline matrix. The minerals can be quartz, feldspars, amphiboles, pyroxenes, and/or olivine. Small round crystallization spheres (= varioles, spherulites) or almond-shaped "amygdales" (both empty and filled with minerals) can also be present, both mm to cm in size, structured in a radiating arrangement or concentrically. → **Volcanic rocks** and **special** cases → 19 / 108

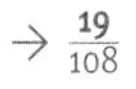

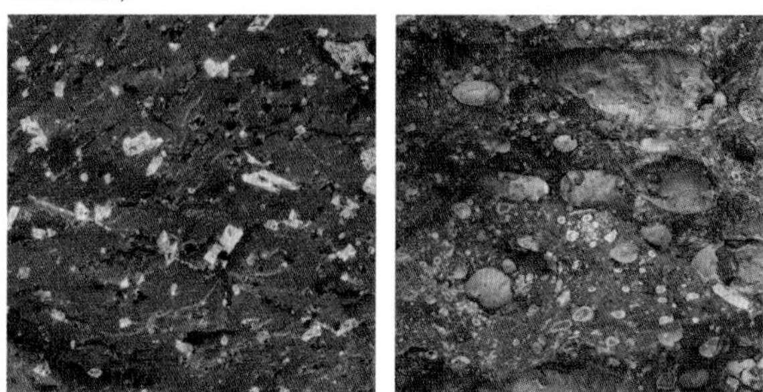

5

Ch. / page

f The rock exhibits one or more kinds of minerals, which form larger crystals/grains and are set in a significantly finer, but still macrocrystalline matrix. → **Macrocrystalline rocks** → 14 / 92

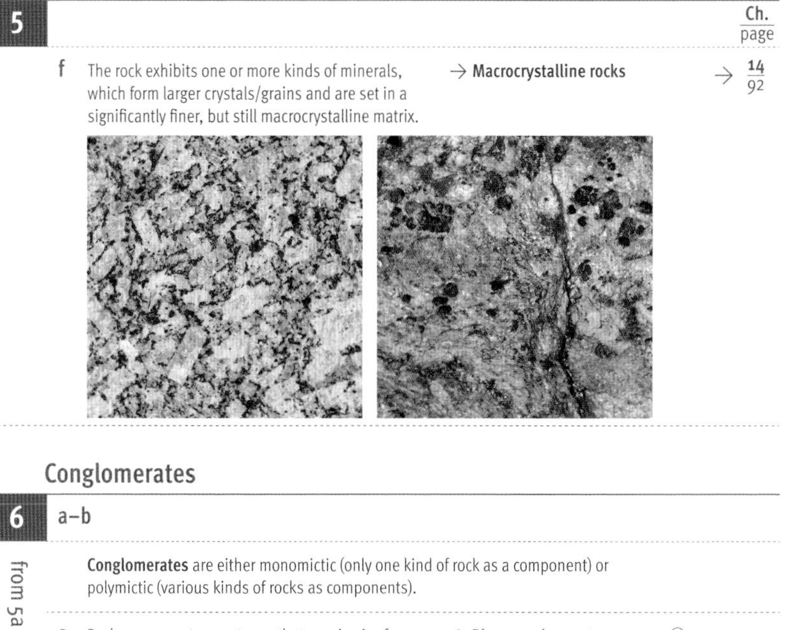

Conglomerates

6 a–b

from 5a

Conglomerates are either monomictic (only one kind of rock as a component) or polymictic (various kinds of rocks as components).

a Rock components are stones that vary in size from a couple of cm to 10–20 cm. → **River conglomerate** ⊕ ⊖

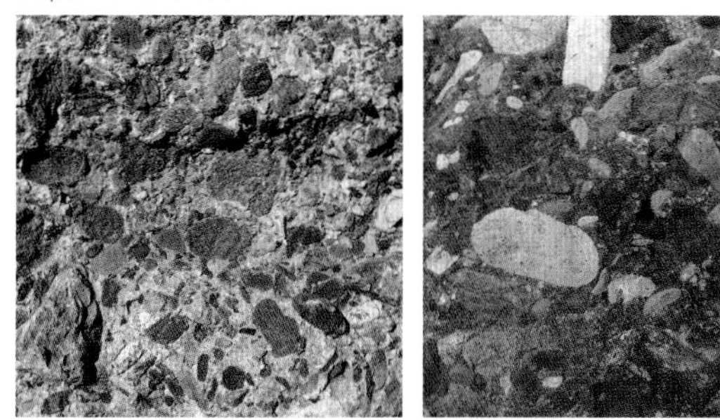

b Rock components are large (starting from sizes in the 10s of dm). → **Beach conglomerate** ⊕ ⊖

Breccias

7 a–g — Ch./page

from 5b

Breccias: can be either monomictic or polymictic.
The demarcation between (a) and (b) is not always easy or unambiguous. Some submarine breccia formations are classified as sedimentary even though they originate in submarine-synsedimentary tectonics.

a Geological context: in a sedimentary environment (for instance as part of a strata series, the filling of a joint or pocket). → **Sedimentary breccia** ⊕ ⊖

Sedimentary breccias can be fine or medium grained to large grained. Limestone or dolomitic breccias are common.

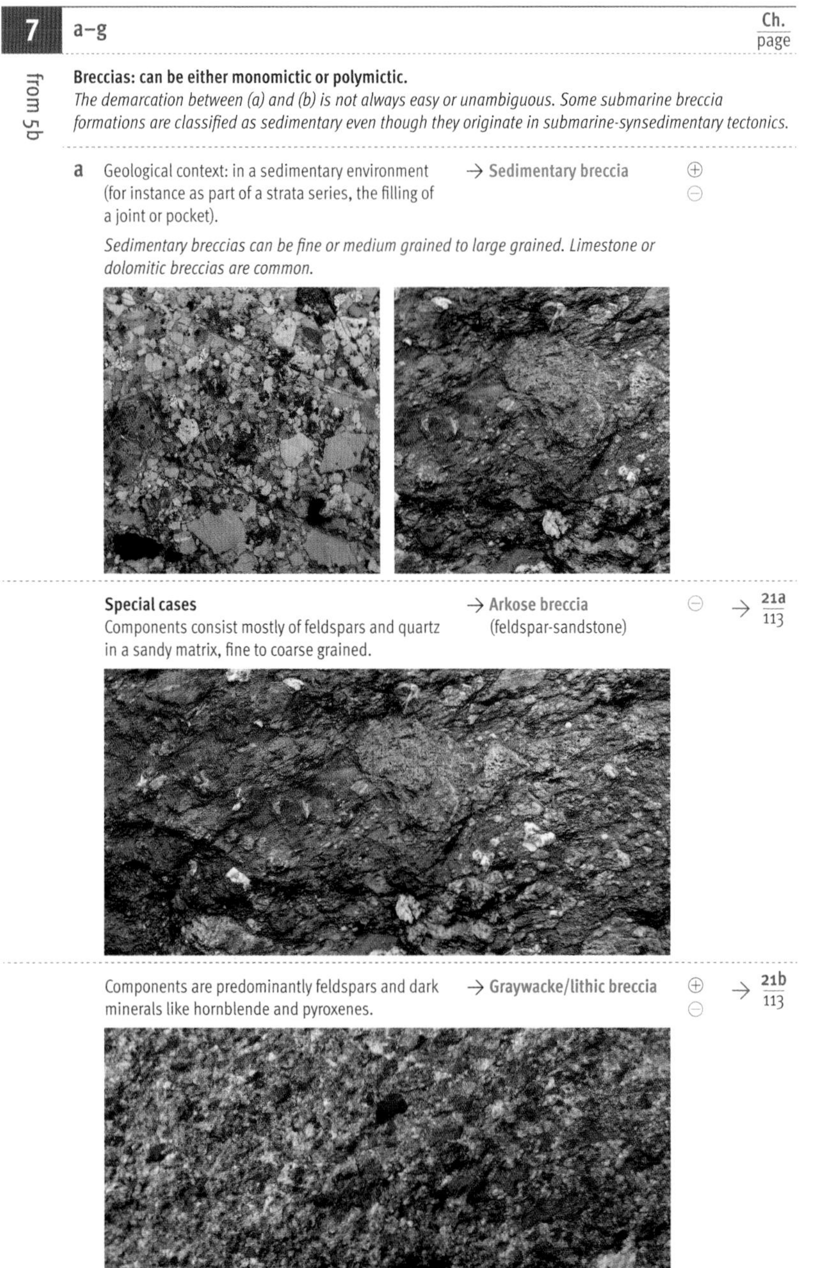

Special cases
Components consist mostly of feldspars and quartz in a sandy matrix, fine to coarse grained. → **Arkose breccia** (feldspar-sandstone) ⊖ → 21a/113

Components are predominantly feldspars and dark minerals like hornblende and pyroxenes. → **Graywacke/lithic breccia** ⊕ ⊖ → 21b/113

7 | Ch./page

b Geological context: in a tectonic environment (in fracture zones, disturbed areas, etc.). → **Tectonic breccia** or **cataclasite** ⊕ ⊖

Unconsolidated tectonic breccias = fault gouge (44h)

c As a defined stratum in fine-grained shallow-water sedimentary rocks → **Storm breccia (tempestite)** ⊕ ⊖

Tempestites can also consist of amassed fossil pieces. Tempestites are encountered quite frequently in the intertidal dolomite strata in the Alps.

→ **Lumachelles** ⊕ → 17c/97

Lumachelles are strata-bound agglomerations of fossil pieces; they develop in high-energy deposition spaces and can also document storm events (a special case of tempestites).

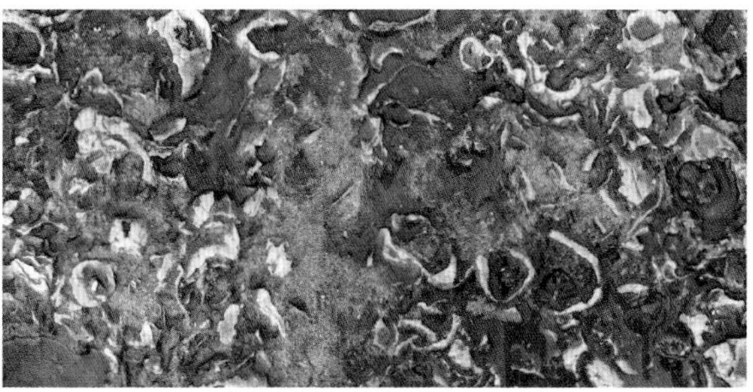

continued overleaf

Breccias

d In a tectonic or subvolcanic context (fracture zones, disturbed areas, etc.) interpenetrated with metalliferous minerals. → **Hydrothermal ore breccia** ⊕ ⊖

A more precise identification according to the specific minerals in question.

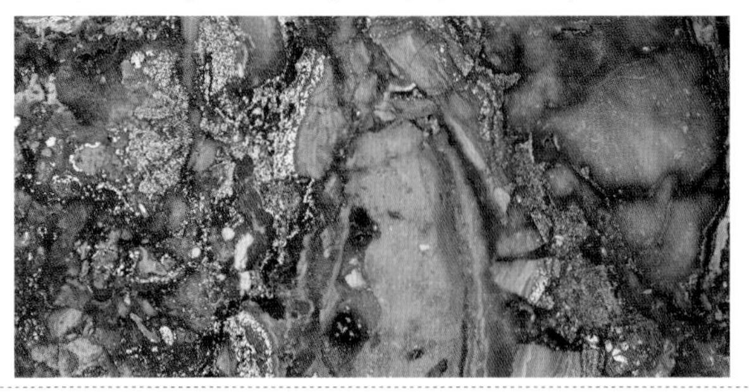

e In a volcanic context: → **Volcanic breccia, tuff breccia** ⊖

It is best to identify larger components as individual rocks! Some volcano-related breccias form from lahars descending from the volcano flanks quite far into the surrounding area ⇢ Volcanosediementary breccia/fanglomerates

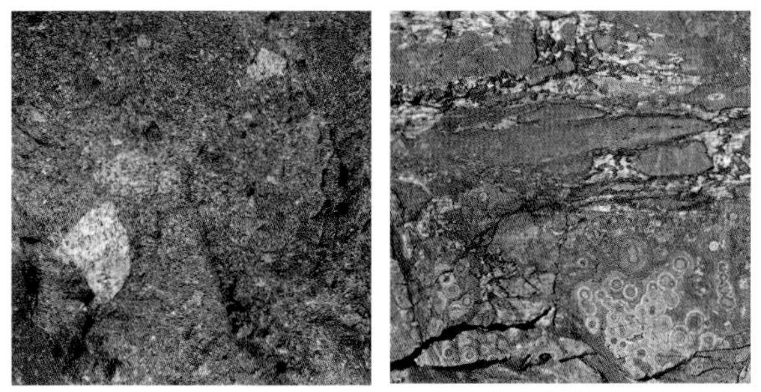

f In the context of plutonic rocks: → **Igneous/magmatic breccia** ⊖

Identify the rocks in question as individual rocks! These kinds of igneous mixed rocks can form because of processes in the magma chamber, starting from a partially solidified state.

7

g Breccias with a black, glass-like matrix: → **Pseudotachylite,** a special case of a tectonic breccia (see also 41c) ⊖

Because of fast and powerful tectonic movements (earthquakes), certain rock parts can be locally molten.

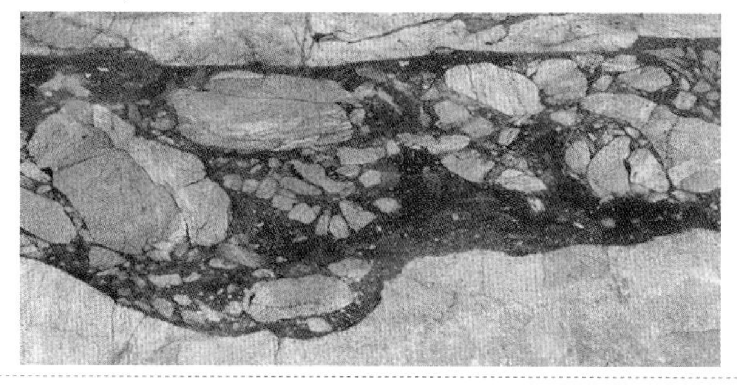

Microcrystalline rocks

8 a–c

		Ch. / page

Grains/minerals are **not** really recognizable with the naked eye or the magnifying glass; the rock is very fine grained, but not glassy.
If you do not come to a satisfactory result with this identification pathway, go back to 2 and choose option 2a.

from 2b

a The rock is very soft (can be scratched with a fingernail, hardness 1–2). → 9 / 78

b The rock is moderately hard (can easily be scratched with a steel blade, hardness 3–5). → 10 / 82

c The rock is hard to very hard (cannot be scratched, or can be scratched only barely with a steel blade, hardness › 6). → 13 / 88

Soft microcrystalline compact rocks

9 a–g

from 8a

a White or very light, chalklike: → **Chalk** (variety of limestone) ⊕

Limestone chalk rocks consist of loosely bound calcareous shells belonging to microfossils in the coccolith family—as in the chalk cliffs of England and northern Germany.

b White, compact, but very soft (can be scratched with a fingernail): → **Gypsum** (microcrystalline) (see also 16a) ⊖

Gypsum is a chemical sediment deposited in the tidal areas of flat, hot, coastal marine environments. It is normally found with salt and dolomite rocks.

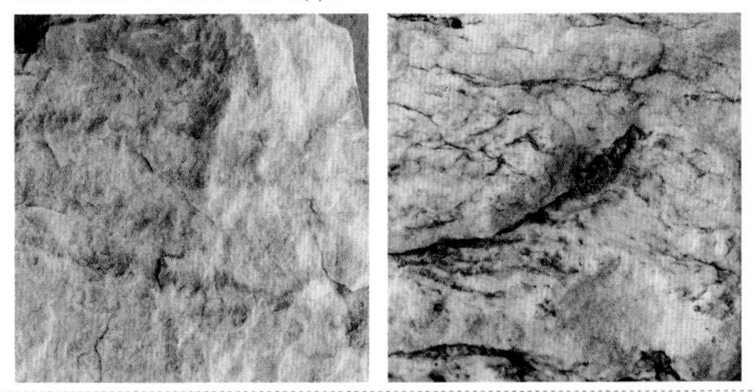

9

C Gray to almost black, sometimes also colored a rust brown; if ground when wet it is argillaceous loamy: → **Mudrock** ⊖

If exhibiting a noticeable to strong reaction with HCl: → **Calcareous mudrock** to **marl** ⊕

continued overleaf

Soft microcrystalline compact rocks

9

d Gray to almost black, sometimes also greenish or red brown/reddish, with a fine cleavage:

→ **Clay shale** or ⊕

→ **Roof slate** (weakly metamorphic mudrock; see also 11a, 29a) ⊖

→ **Black shale/oil shale** (clay shale with noticeable to high content of organic substances) ⊖

9

Slightly harder consistency, less argillaceous, exhibiting a noticeable reaction to dilute HCl: → **Marl slate** (weakly metamorphic marl, with carbonate; see also 11b) ⊕

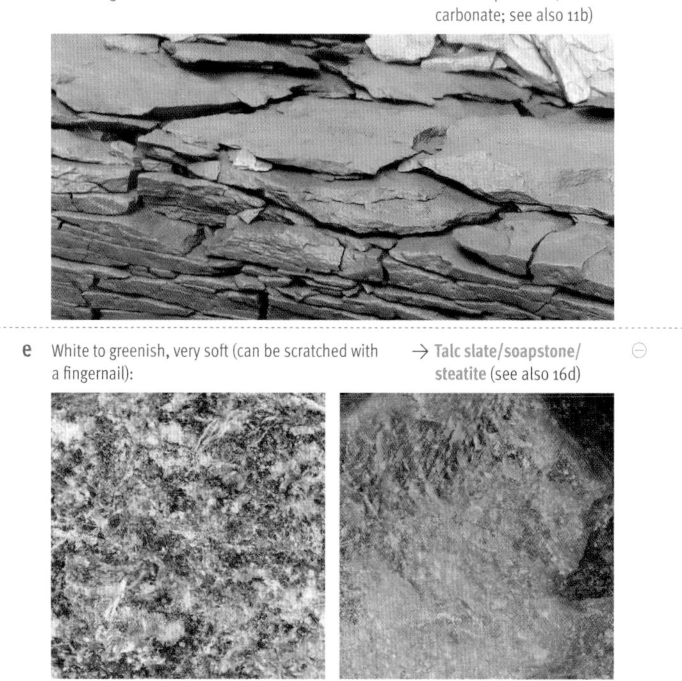

e White to greenish, very soft (can be scratched with a fingernail): → **Talc slate/soapstone/ steatite** (see also 16d) ⊖

f Medium to dark brown, sliceable to an extent, also exhibits a distinctive oily smell to some degree: → **Lignite** (brown or bituminous coal) ⊖

continued overleaf

Soft microcrystalline compact rocks

9

g Rust brown or any possible color from creamy to dark brown. It can be heterogeneous and contain small spheres and round concretions composed of iron oxides/hydroxides. → **Bauxite (left)**

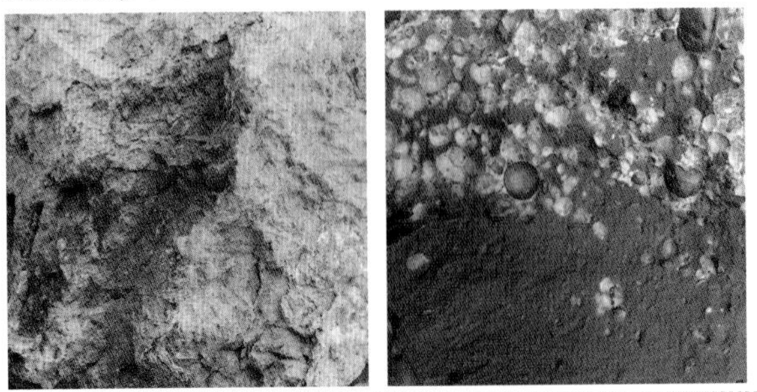

Moderately hard microcrystalline compact rocks

10 a–b

Ch./page

from 8b

a Massive, not schistose at the hand specimen level (but can still exhibit fine bands). → **12**/85

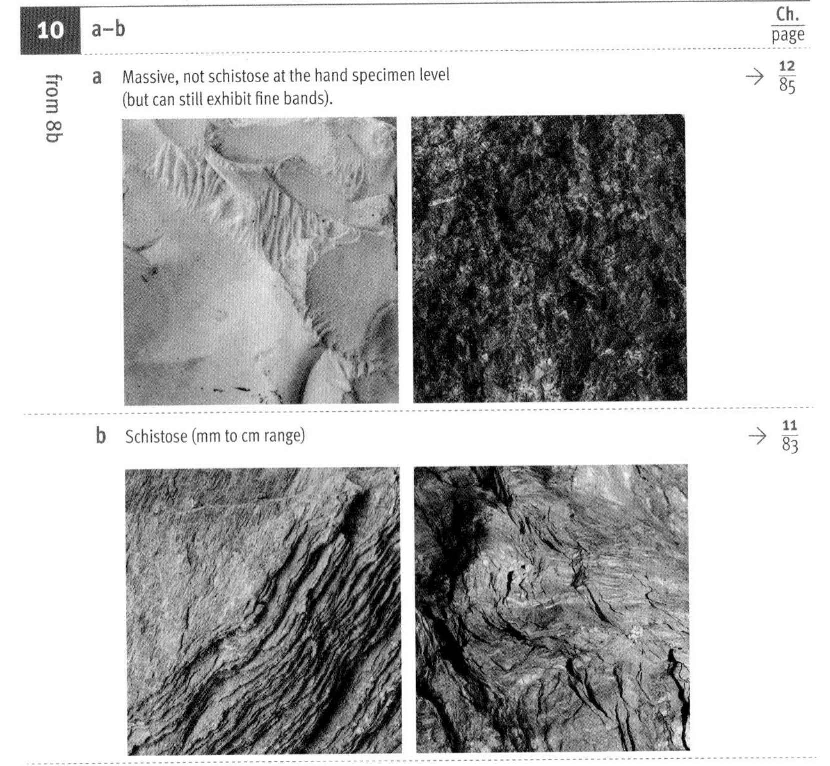

b Schistose (mm to cm range) → **11**/83

Schistose and soft microcrystalline compact rocks

11 a–e

from 10b

a Gray to black, very finely schistose: → **Clay slate/roof slate** (see also 9d, 29a) ⊖

Roof slate is a weakly metamorphic mudrock.

b Gray to black, somewhat harder, with very fine grainy-crystalline layers between cleavage surfaces; occasionally also very fine alternate layerings of marl/clay slate: → **Marl slate, (calcareous) shale** (see also 9d, 29a) ⊕

c Gray to black, soft, with a slight silky gloss on the cleavage surfaces, sometimes already recognizable as micaceous minerals: → **Phyllite** (left) to **sericite slate** (right) (see also 12b) ⊖

continued overleaf

Schistose and soft microcrystalline compact rocks

11

d Greenish to clearly green, partly also spotty with light areas rich in mica and green areas rich in chlorite: → **Chlorite slate/chlorite-sericite-phyllite** (see also 29b) ⊖

e Light green to black green, the colors often irregularly divided; sometimes with shiny and/or striated slickenside marks, in between also massive and blackish; often contains black ore mineral grains (mostly magnetite): → **Serpentinite/serpentine schist** (see also serpentinite at 12d, 19f, and 42a) ⊖

Nonschistose, moderately hard, microcrystalline compact rocks

12 a–e

from 10a

a Variable colors, from almost white to light to dark gray to almost black, also yellowish, brownish, or reddish: → **Micritic/very fine-grained limestone** ⊕

or: → **Micritic/very fine-grained dolomite** ⊖

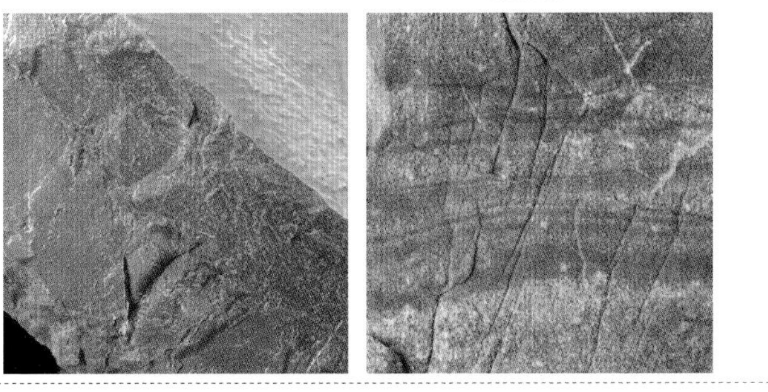

Very soft and smudges when wet (in other words, argillaceous): → **Marly limestone/dolomite** ⊕

continued overleaf

Nonschistose, moderately hard, microcrystalline compact rocks

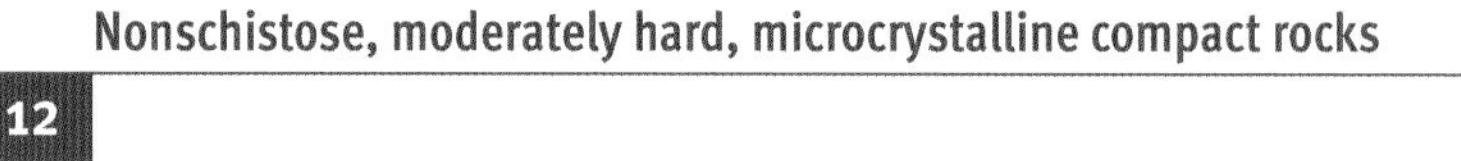

b Rocks with very fine, strictly parallel banding; rocks are dense, crystalline: → **Calcareous mylonite** ⊕

Calcareous mylonites occur in major tectonic shear zones in limestone series (e.g., nappe thrusts).

c Medium to dark green, massive: → **Greenstone** ⊕
(see also 23d, 29e) ⊖

Weakly to moderately metamorphic basalt without cleavage.

d Dark green to blackish, also partially exhibits green to light green surfaces, sometimes with enclosed gold-colored shiny crystals (= pyroxenes): → **Massive serpentinite** ⊖
(see also 11d, 19f) ⊖

All transitions to strongly schistose serpentinite are possible (11e). Often contains black ore mineral grains (mostly magnetite). Serpentinites can also exhibit very glossy and grooved slickenside surfaces or extreme veining (with vein fillings consisting of calcite, serpentine, tremolite, asbestos, calcite, etc.; see also 42a).

12

d *continued ...* → **Massive serpentinite** (see also 11d, 19f) ⊖ ⊖

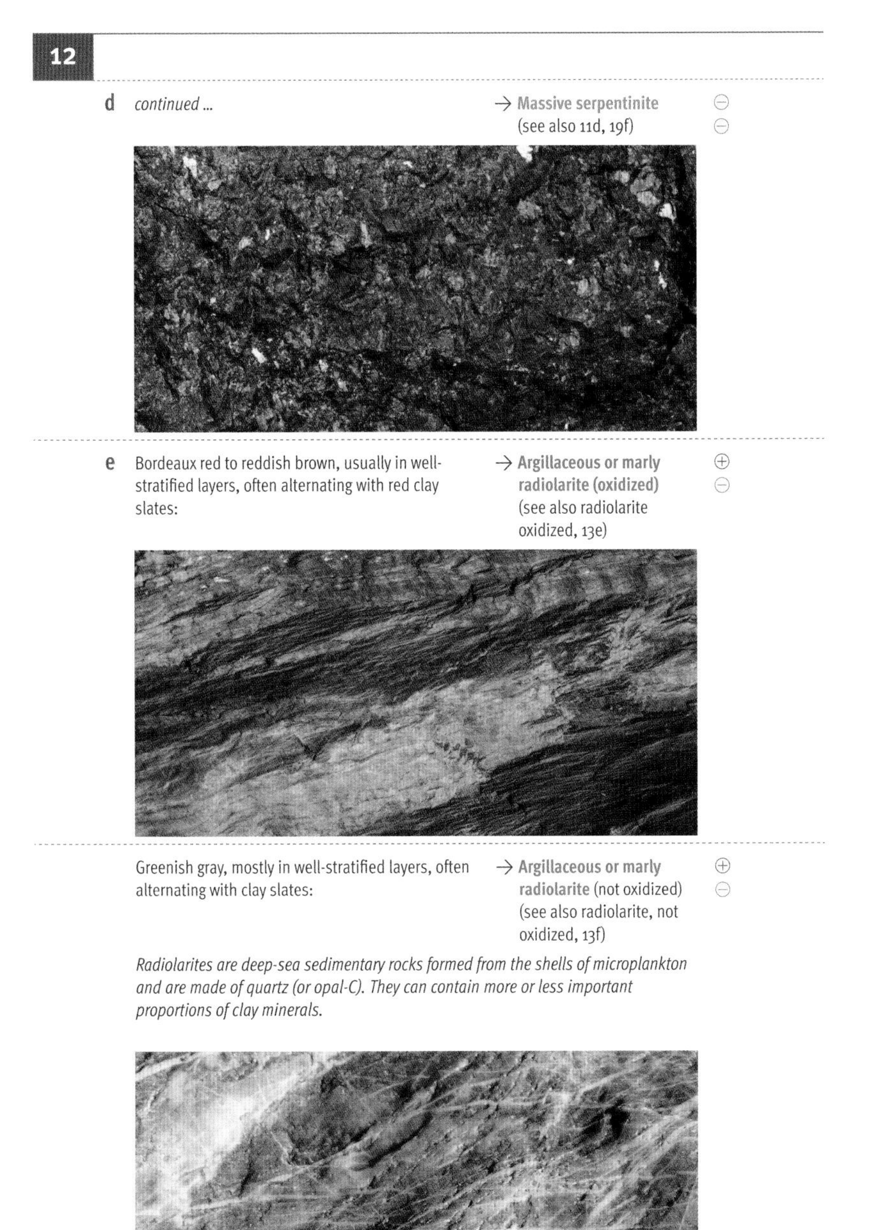

e Bordeaux red to reddish brown, usually in well-stratified layers, often alternating with red clay slates: → **Argillaceous or marly radiolarite (oxidized)** (see also radiolarite oxidized, 13e) ⊕ ⊖

Greenish gray, mostly in well-stratified layers, often alternating with clay slates: → **Argillaceous or marly radiolarite** (not oxidized) (see also radiolarite, not oxidized, 13f) ⊕ ⊖

Radiolarites are deep-sea sedimentary rocks formed from the shells of microplankton and are made of quartz (or opal-C). They can contain more or less important proportions of clay minerals.

Hard microcrystalline compact rocks

13 a–h

from 8c

a The rock is black to dark gray, occasionally with individual disseminated crystals: basaltic or andesitic volcanics.

1. Disseminated crystals are feldspar or hornblende: → **Andesite** (see 4d, 19c) ⊖

2. Disseminated crystals are augite and/or olivine: → **Basalt** (see 4d, 19d) ⊖

Occasionally "amygdales" (former volcanic gas bubbles now filled with minerals).

b Matrix is dark brown to dark blackish green, and disseminated minerals are often black hornblende and/or dark brown biotite: → **Lamprophyre** ⊕ → **always appear as dikes!** ⊖ (see also 19e)

Lamprophyre dikes quite common in many plutonic complexes. They can be distinguished from basalt/andesite mainly because of their higher potassium content. An extended series of lamprophyre types can be distinguished (minette, vogesite, kersantite, spessartite, etc.). They can be distinguished from basalt/andesite because they appear in dikes and because of their biotite content and brownish weathering coloration, among other differences.

13

C The rock is light colored, whitish, gray, reddish, or rust brown, often with disseminated crystals: **acid volcanics**.

1. Disseminated crystals are quartz and/or potassium feldspar. Rhyolites (left) from ash flow deposits that exhibit ragged and flame-like structures = ignimbrite (also welded tuff) (right). → **Rhyolite/ignimbrite** (see also 4e, 19a) ⊖

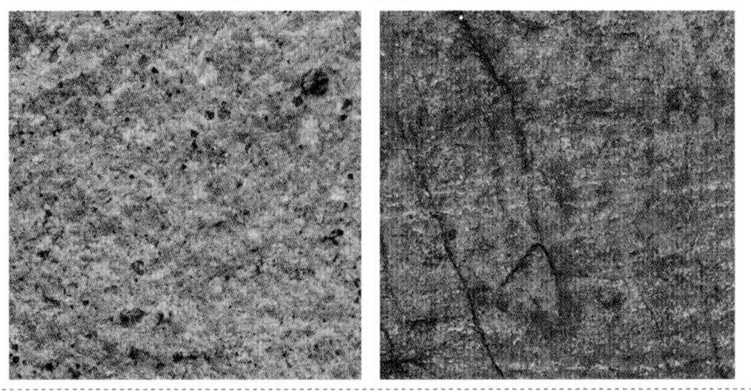

2. Disseminated crystals are feldspar: 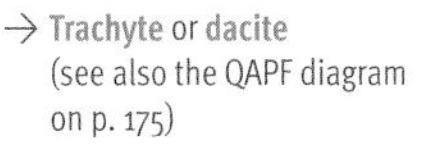 → **Trachyte** or **dacite** (see also the QAPF diagram on p. 175) ⊖

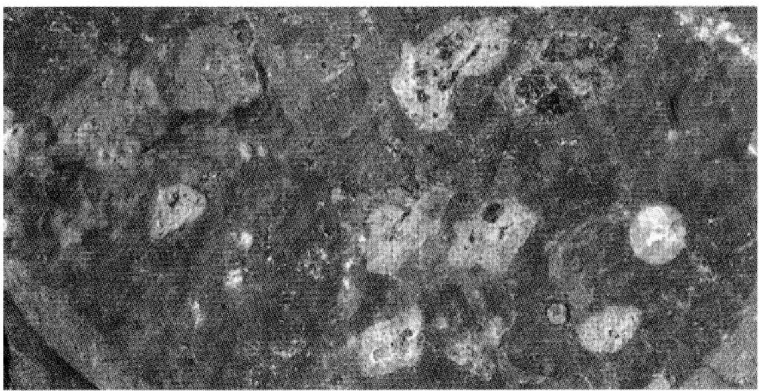

3. Disseminated crystals are foids (feldspar representatives, for instance nepheline, leucite): → **Feldspathoid**-containing or "foid" volcanics (phonolite, leucite-tephrite, etc.) (see also the QAPF diagram on p. 175) ⊖

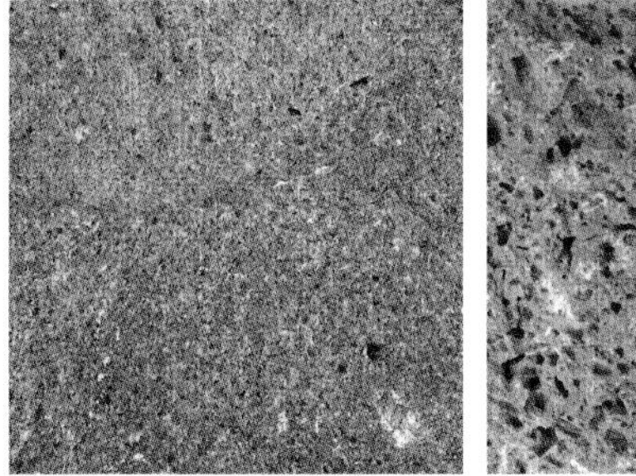

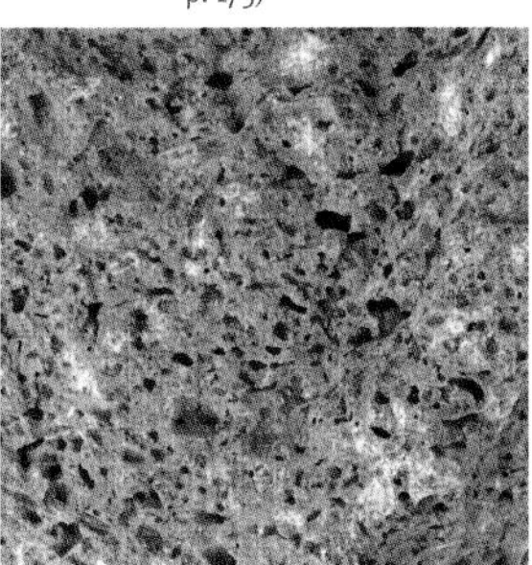

continued overleaf

Hard microcrystalline compact rocks

13

d The rock is brown to dark brown or blackish gray, sometimes with a slight purple tint; often very tough, splintery: → **Hornfels** ⊕ ⊖

Strong contact metamorphic mudrock (pelite). Can definitely be confused with basalt, but its great hardness and splintery nature, as well as the geological context, should enable a correct classification.

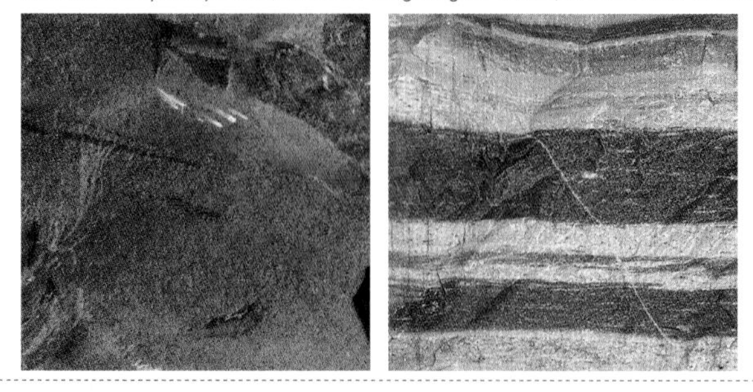

e The rock is Bordeaux red to reddish brown, usually in well-stratified layers, often alternating with red clay slate: → **Radiolarite** (oxidized) or **lydite** ⊖

Red/green radiolarites are known above all from ocean crust remnants (ophiolites) in younger mountain ranges. Paleozoic radiolarites are usually gray to black and are referred to as lydite. As far as origins are concerned, however, they are identical.

f The rock is light green to gray green, usually in well-stratified layers, often alternating with red clay slate: → **Radiolarite** (nonoxidized) or **lydite** ⊖

The same observations hold as for 13e.

13

g The rock is mostly gray to black, or brownish reddish, in nodules, mostly with a thin whitish outer layer: → **Flint, chert (silex)** ⊖

Silex nodules are frequently present in fine-grained limestones, usually having accumulated in layers. They originate from smaller parts of plankton shells made of quartz (or opal-CT) that dissolve during rock consolidation. The quartz that has been dissolved in interstitial water is later precipitated at specific horizons in the form of hard, tough, microcrystalline nodules. Because of their high resistance to weathering, they are often "weathered out" of the limestone.

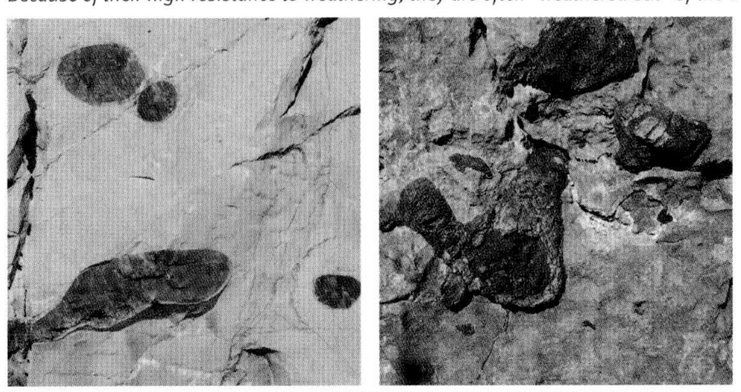

h The rock—varying in coloration—is very finely banded and can contain extensively drawn-out structures: → **Mylonite** ⊕ ⊖

Mylonites can be formed from the most diverse rocks by extreme shear deformations, for instance at the base of thrust faults. Calcareous mylonites are a special case (see 12b).

continued overleaf

Macrocrystalline rocks

14 a–d — Ch. / page

Grains/minerals are recognizable with the naked eye and/or with a magnifying glass, even when in certain circumstances they can be very small.

from 2a and 5f

a The rock consists for the most part, or entirely, of **only one kind of mineral** (monomineralic) or component type; these can also be spherical aggregates.

Examples:
peridotite (left)
oolite-limestone (right)

→ 15 / 93

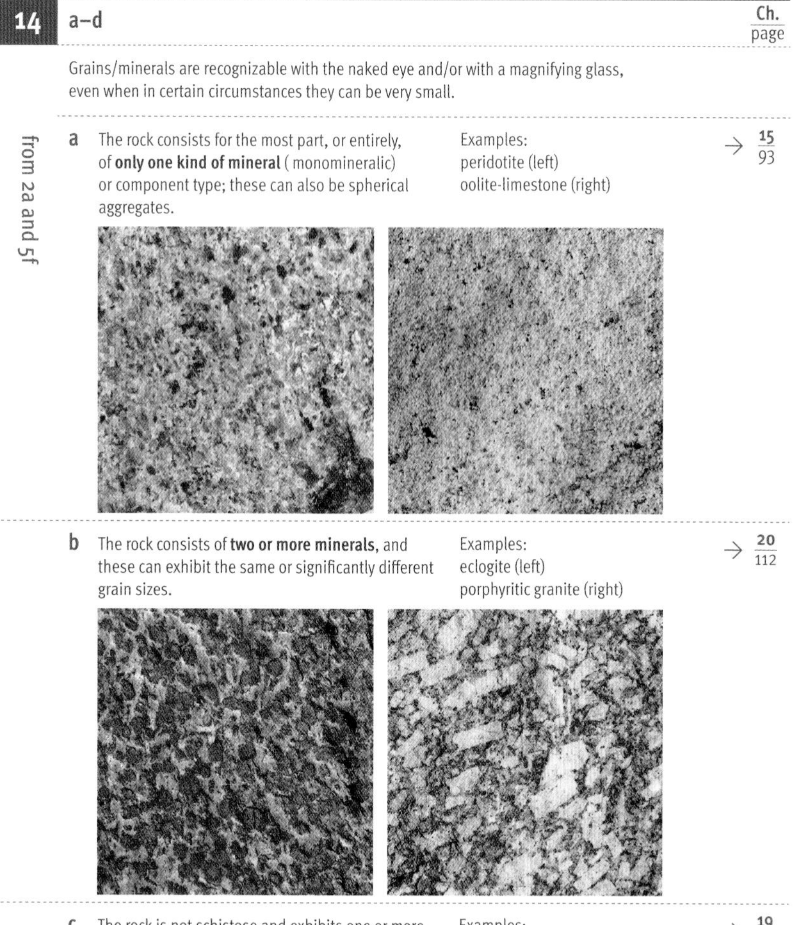

b The rock consists of **two or more minerals,** and these can exhibit the same or significantly different grain sizes.

Examples:
eclogite (left)
porphyritic granite (right)

→ 20 / 112

c The rock is not schistose and exhibits one or more recognizable kinds of minerals in a clearly finer, mostly very fine-grained matrix. The minerals can present more or less idiomorphic forms (crystal shapes).

Examples:
andesite (left)
rhyolite (right)

→ 19 / 108

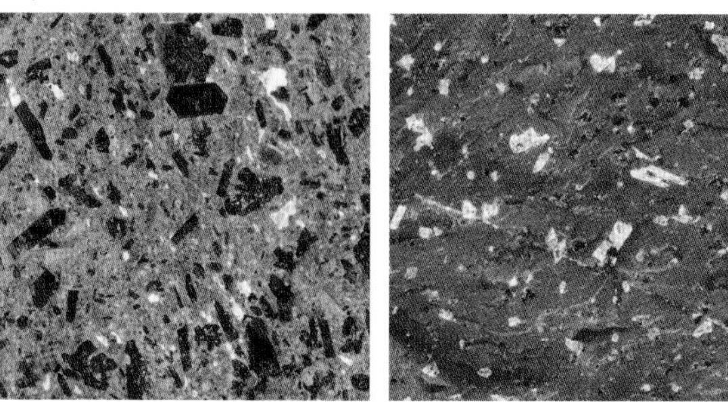

14 — Ch. / page

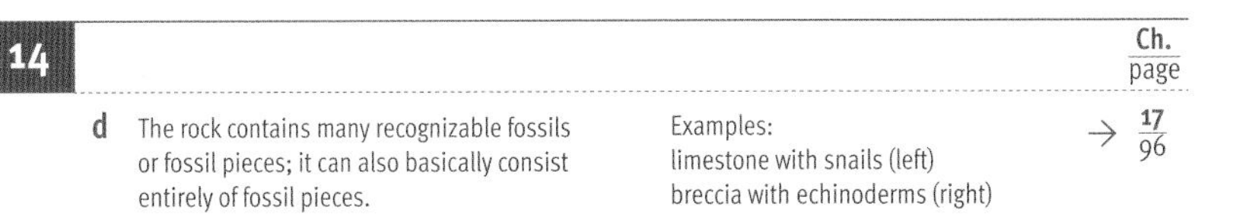

d The rock contains many recognizable fossils or fossil pieces; it can also basically consist entirely of fossil pieces.

Examples:
limestone with snails (left)
breccia with echinoderms (right)

→ 17/96

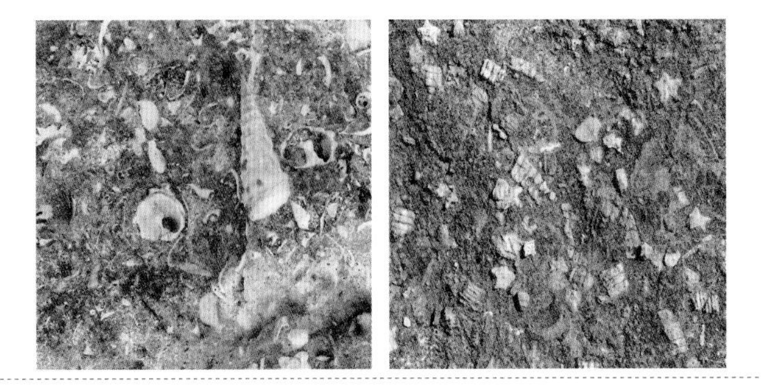

Monomineralic macrocrystalline rocks

15 a–d — Ch. / page

from 14a

Testing the hardness of rock-forming minerals
Beware of the **“sandstone trap”**! → Especially in the case of fine-grained minerals, do not only scratch the grains with the test blade, but also scratch a flat surface of the testing material with the rock to ascertain whether there are scratches.

a The rock or the rock-forming mineral is very soft to soft (can be scratched with a fingernail, hardness 1–2). → 16/94

b The rock or the rock-forming mineral is moderately hard (is not scratchable with a fingernail but can be very easily or easily scratched with a steel blade, hardness 3–5). → 17/96

c The rock or the rock-forming mineral is hard or very hard (barely scratchable or not scratchable with steel, hardness › 6). → 18/102

d The rock or the rock-forming mineral has a metallic ore character and exhibits an elevated specific weight; ore. → 40/148

Soft monomineralic macrocrystalline rocks

16 a–e

from 15a

a The rock is white to beige, grainy crystalline: → **Gypsum** (see also 9b) ⊖

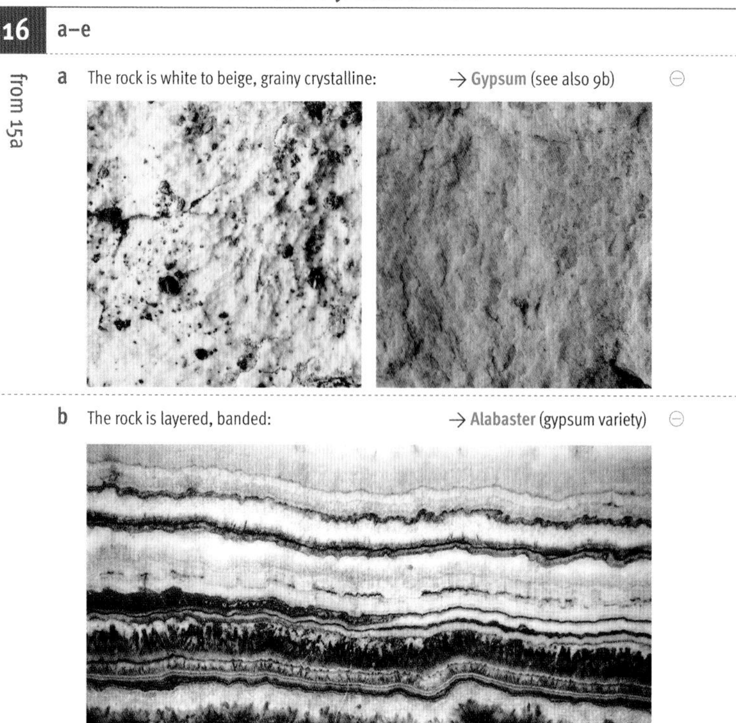

b The rock is layered, banded: → **Alabaster** (gypsum variety) ⊖

c The rock is crystalline and tastes salty: → **Rock salt** ⊖

Rock salt is the most important evaporitic rock for humans; it was formed on flat, hot seacoasts often together with gypsum and dolomite.

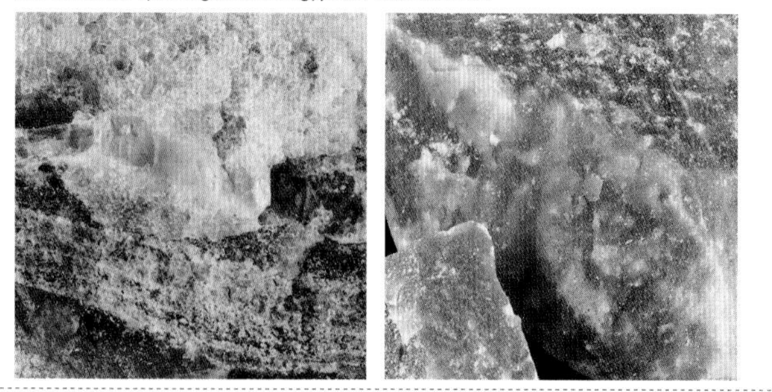

16

d The rock is white to light green, often flaky, feels soapy: → **Soapstone/Steatite** (talc rock) (see also 9c)

Soapstones/steatites (talc rocks) are formed at and around bodies of serpentinite as reaction products with the neighboring rocks.

e The rock is black to blackish gray, with an oily luster, greasy: → **Graphite**

Graphite is the higher metamorphic equivalent of black coal. In many metamorphic sedimentary rocks it leads to a gray coloration.

Moderately hard monomineralic macrocrystalline rocks

17 a–i

from 14d and 15b

a The rock is fine to medium grained; the colors vary from light gray to black, but yellowish to brownish is also possible. It consists of very small to small spathic crystals (sparkling because of good fissility)

→ **Limestone** finely to roughly sparry (crystalline) ⊕

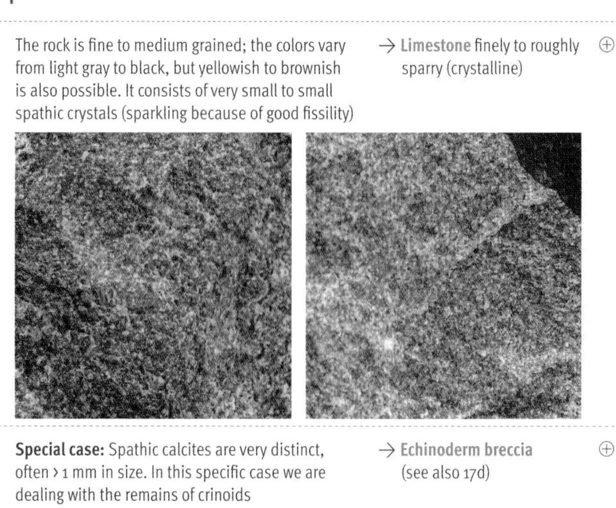

Special case: Spathic calcites are very distinct, often > 1 mm in size. In this specific case we are dealing with the remains of crinoids

→ **Echinoderm breccia** (see also 17d) ⊕

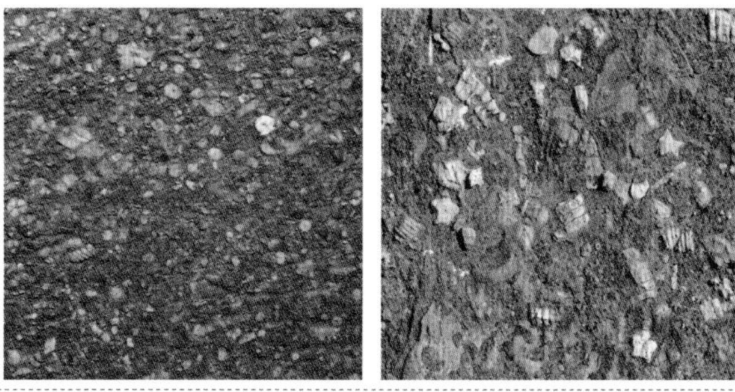

b The rock is fine to medium grained; color varies from pure white to gray, but other tones are also possible (yellowish, ocher, pink, bluish). The rock is composed of clearly recognizable crystalline-spathic crystals and can contain minor amounts of other minerals, e.g., mica:

→ **Calcite or dolomitic marble** (differentiation is possible only with the HCl test) ⊕ ⊖

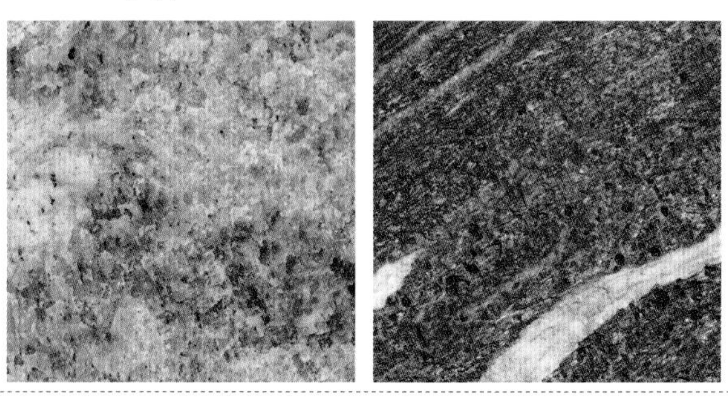

17

Special case: The grains are poorly connected to one another; the rock has a sandy-grainy character: → **Sugary dolomite marble** ⊕ ⊖

To date it is not completely clear why these sugar-grained dolomitic marbles exhibit such poor cohesiveness.

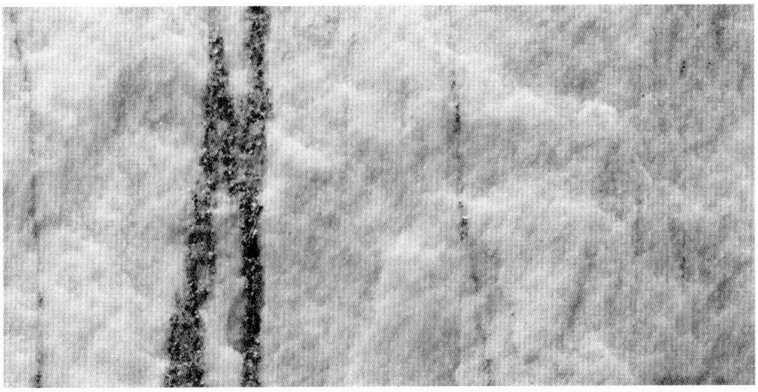

from 5d

C The rock is fine to medium grained and consists mostly of up to ca. 2 mm round components, which often exhibit a shelled internal structure. They resemble fish roe → **Oolite** or **oolitic limestone** ⊕

Ooids are formed in shallow, current-rich tropical oceans via precipitation of calcite needles on algae in the constant movement of ocean currents.

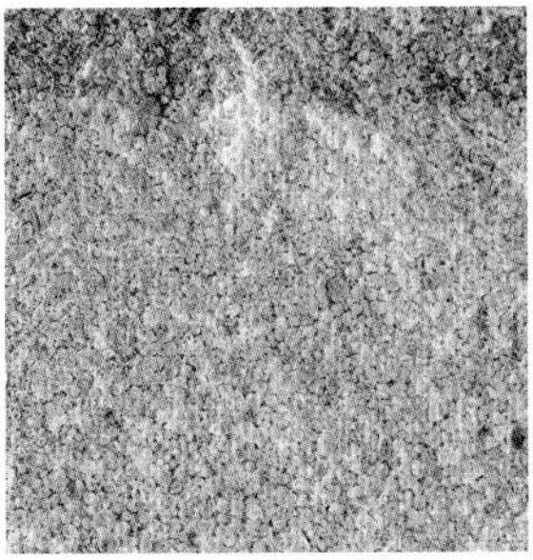

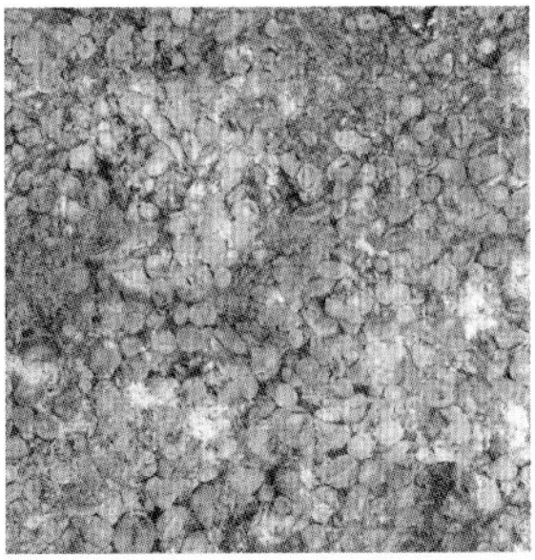

Concretions are larger (up to > 1 cm), frequently irregularly shaped (= oncoids): → **Oncolite/oncolitic limestone** ⊕

Oncolites are structurally analogous to ooids.

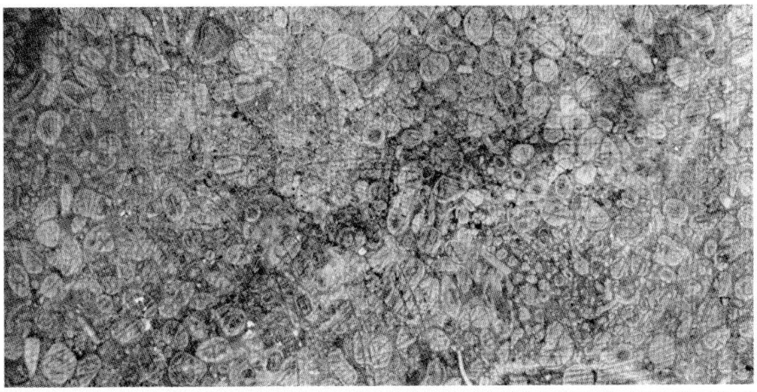

continued overleaf

Moderately hard monomineralic macrocrystalline rocks

17

from 5c

d The rock is fine to coarse grained and consists mostly of either pieces or entire shell remains of marine life forms; the matrix can be either micro- or macrocrystalline → **Bioclastic limestone** ⊕

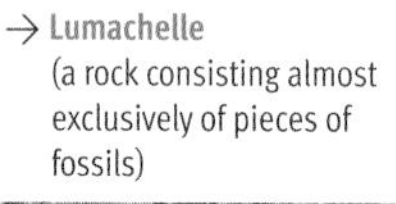

→ **Lumachelle** ⊕
(a rock consisting almost exclusively of pieces of fossils)

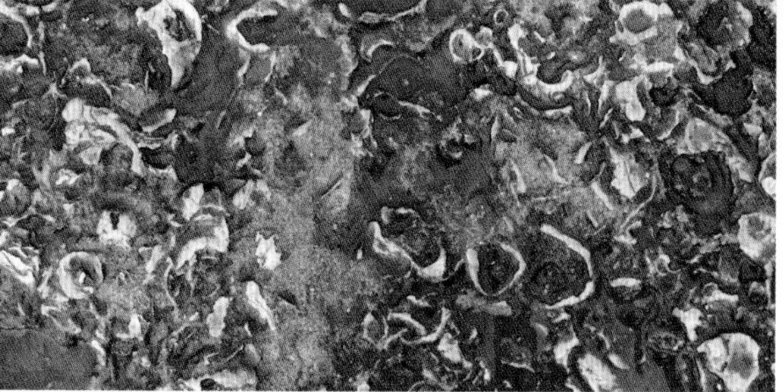

Special cases: Fossil components (bioclasts) are → **Fossiliferous limestone** containing mussel shells ⊕

Mussels/mussel remains:

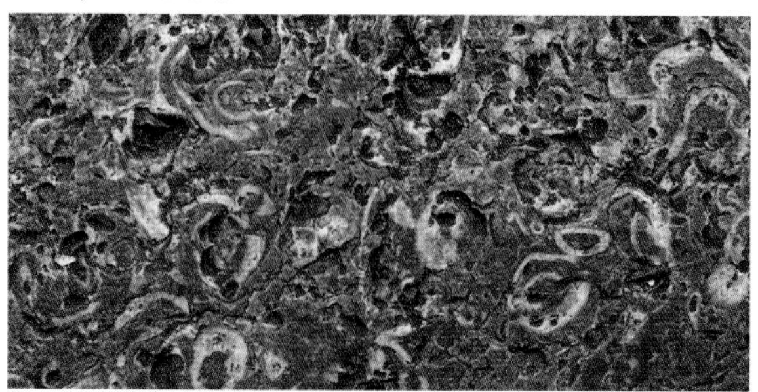

17

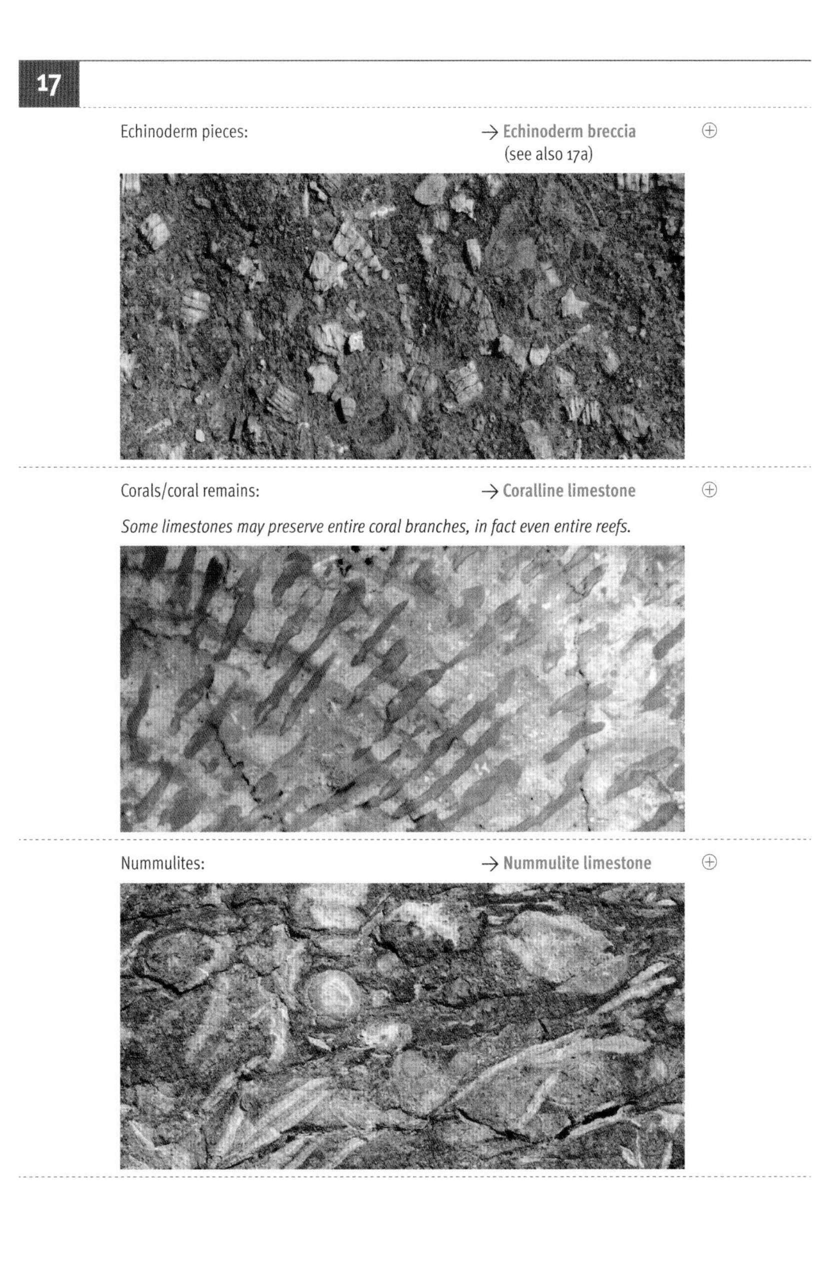

Echinoderm pieces: → **Echinoderm breccia** (see also 17a) ⊕

Corals/coral remains: → **Coralline limestone** ⊕

Some limestones may preserve entire coral branches, in fact even entire reefs.

Nummulites: → **Nummulite limestone** ⊕

continued overleaf

Moderately hard monomineralic macrocrystalline rocks

17 | Ch./page

e The rock is fine to medium grained, crystalline, colorless to gray (pink to lilac tones are also possible) → **Anhydrite-rock** ⊖

Anhydrite is a metamorphic gypsum stone—if you add water you will once again obtain gypsum.

f The rock is pitch black, either grainy matte: → **Black coal** (left) ⊖

or crystalline shiny: → **Anthracite** (right) ⊖

g The rock is medium to coarse grained, usually white, but an ocher-brown coloration is also possible; an interlocked mass of individual grains in which you can more or less clearly recognize rhombic cleavage surfaces: → **Calcite** from a joint/vein or a concretion (see 43e) ⊕ → 43e/157

17

h The rock is ocher yellow, consists of calcite crystals with a conspicuous layered structure; the structure can develop in wavy or circular patterns → **Speleothem, dripstone** ⊕

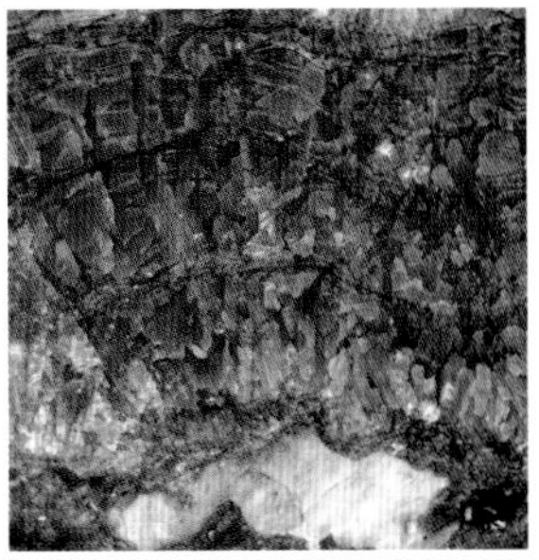

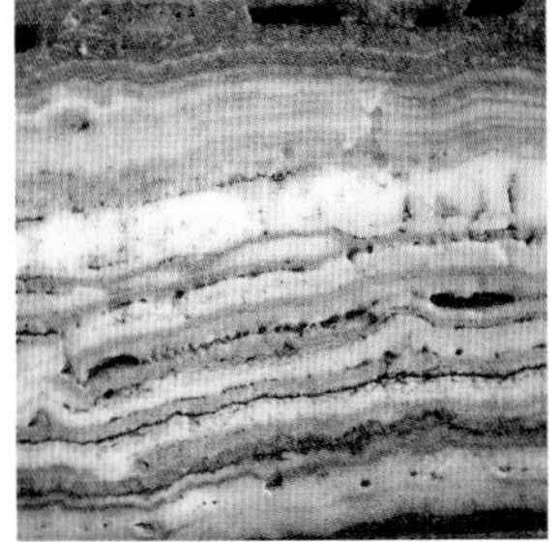

i The rock consists of other moderately hard minerals: for instance baryte or heavy spar (left), fluorite lode (right). These kinds of rocks often form lode or vein fillings near hydrothermal formations or mineralizations. → **Special cases** ⊖

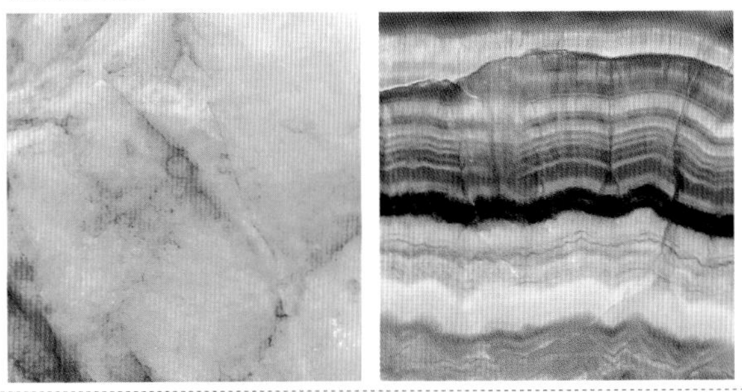

Hard monomineralic macrocrystalline rocks

18 a–g

from 15c

a The rock consists for the most part of more or less well-rounded, more or less compact, interlocked quartz grains (whitish to grayish, very hard, conchoidal fracture, oily luster on fracture surfaces): → **Quartz sandstone** (see also 22b) ⊕ ⊖

The hardness of quartz sandstones is due primarily to the interlocking of the quartz grains, secondarily to the kind of cement.

Cement = calcite (HCl test!)

Soft sandstone, for instance most younger molasse sandstones (greenish gray) ⊕

18

Cement = quartz

Gritstone (hard sandstone), like the Millstone Grit of England or gritstones near Boulder, Colorado ⊖

Quartz sandstones from arid sedimentation conditions (arid basins, deserts) are often red, such as the red sandstones interbedded with sandstones and shales of other colors in the Chinle Formation exposed in the Painted Desert (Arizona, USA).

Quartz sandstones can exhibit stratifications from the mm to the m range. They can also appear as compact massive-directionless rocks for intervals of several meters. River sandstones often exhibit a cross-stratification, while sandstones from desert dunes often exhibit dune patterns.

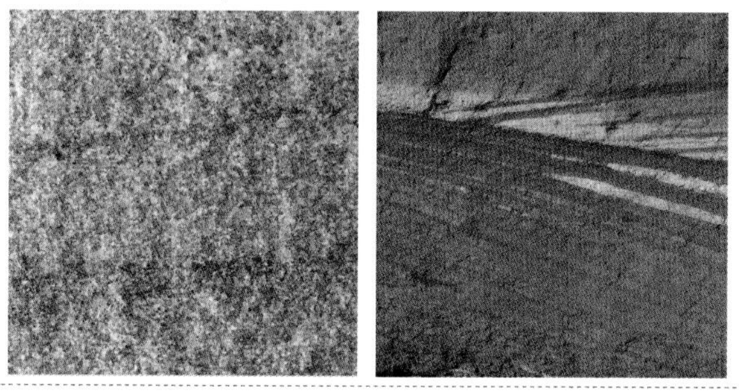

Special case: Greenish sandstone

→ **Green sandstone/ glauconitic sandstone** ⊕ ⊖

Glauconite is a green clay mineral that can form in a reducing environment in flat ocean basins.

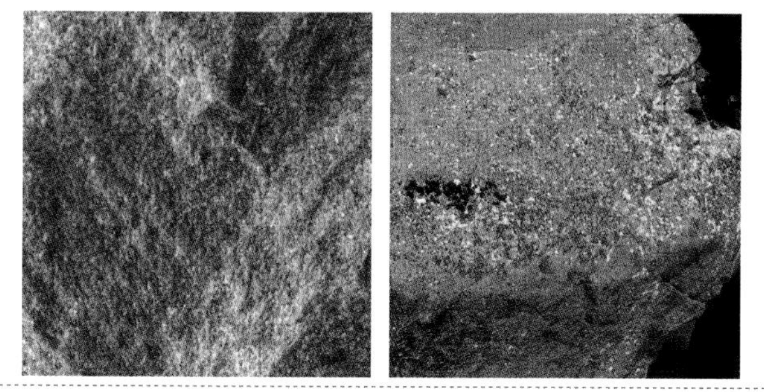

continued overleaf

Hard monomineralic macrocrystalline rocks

18

from oo

b The rock is fine to medium grained and compact but mostly cleaves flatly; whitish to greenish; it consists of very closely interlocked quartz grains, where the original rounded forms of the grains are hardly or no longer recognizable. These rocks also often contain some white mica parallel to the line of stratification (cleavage):

→ **Quartzite** (metamorphic quartz sandstone) ⊖

The demarcation between compact sedimentary quartz sandstone (gritstone) and metamorphic quartzite is fluid.

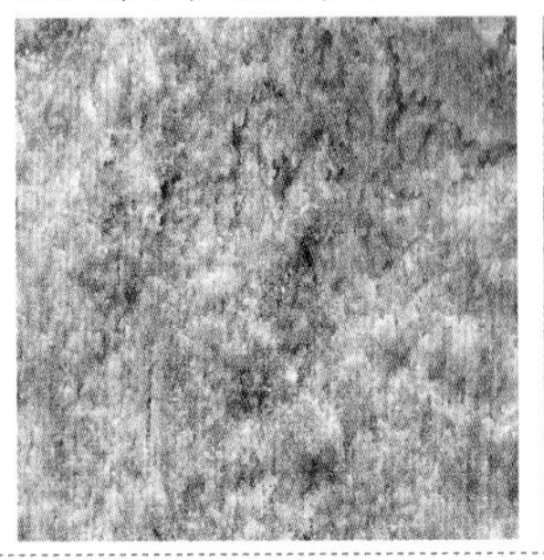

c The rock consists of whitish, also yellowish to rust-red crystals that are barely scratchable with steel. The grains are spathic (good fissility) and mostly exhibit a suture of the twin plane at the grain's midpoint:

→  **Leucosyenite** ⊖

18

Special case:
A special case, which is very well known because of its popularity as natural stone for kitchens, facades, and floors, is the so-called → **Larvikite**

Grayish-blue, 1–5 cm alkali feldspar and black pyroxenes are closely intercrystallized with one another. The alkali feldspars exhibit, also because of internal differentiation, a conspicuous blue play of colors, which gives the rock its special fascination.

d The rock consists almost solely of closely interlocked olive-green grains. The specific weight is comparatively high (around 3.3 g/cm³): → **Peridotite** (mantle rock) (see also 23b, both photos)

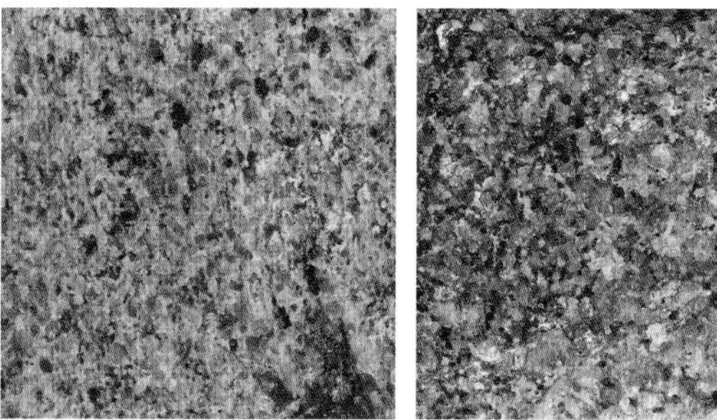

continued overleaf

Hard monomineralic macrocrystalline rocks

18

d *continued ...*

Also, a much rarer case: → **Olivine fels** metamorphic

Other minerals can also be present in smaller quantities, for instance black grains of magnetite or chromite, green pyroxenes, or light brown mica (phlogopite).

In the Alps the peridotites from the area around Ivrea (Finero, Balmuccia, Baldissero) are well known, and they sometimes exhibit metamorphic structures and can contain considerable amounts of the Mg-biotite phlogopite (photo above right).

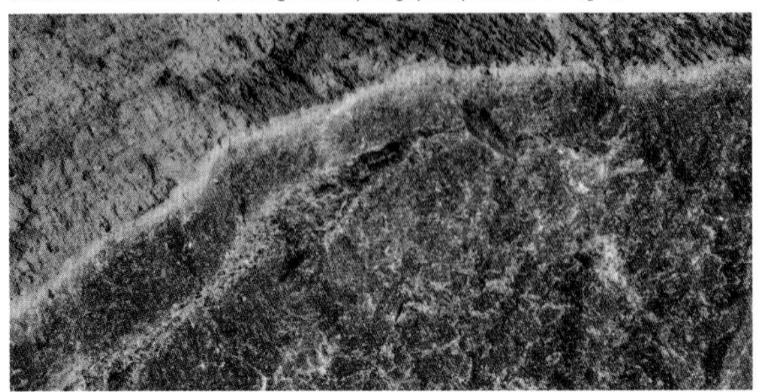

e The rock is composed of medium- to coarse-grained, mutually interlocked white grains. The grains are significantly harder than steel, no fissility

→ **Quartz** from the fillings of hydrothermal joints, veins, or concretions, respectively (see also 43a

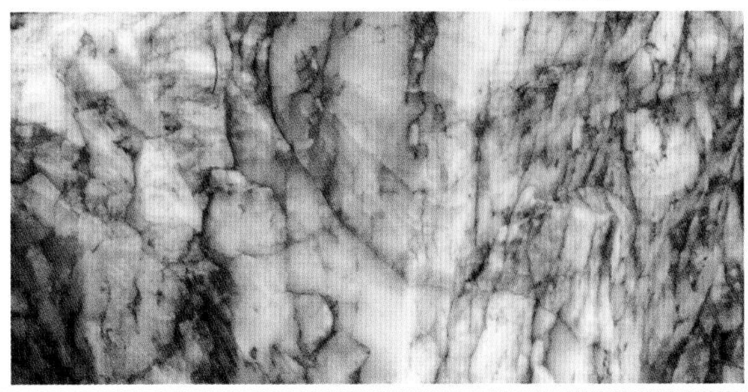

18

The grains are barely scratchable with steel and exhibit cleavage planes: → **Feldspar** (often albite) from the fillings of hydrothermal joints, veins, or concretions, respectively (see also 43b) ⊖

f The rock is medium to coarse grained and consists almost solely of light-colored plagioclase feldspar: → **Anorthosite** (see also 27d) ⊖

Anorthosites are rare and special plutonites that occur mostly in ancient continental shields (for instance in Norway, Finland, Ukraine, and Canada).

g The rock consists of black to blackish-green minerals with rounded to truncate pyramidal shapes that split along 2 longitudinal cleavages following a "small step" pattern, and gleam accordingly. The specific weight is relatively high (around 3.3 g/cm³). The "small-step" cleavage is roughly at right angles (90°), pyroxene minerals: → **Pyroxenite** (see also 23b) ⊖

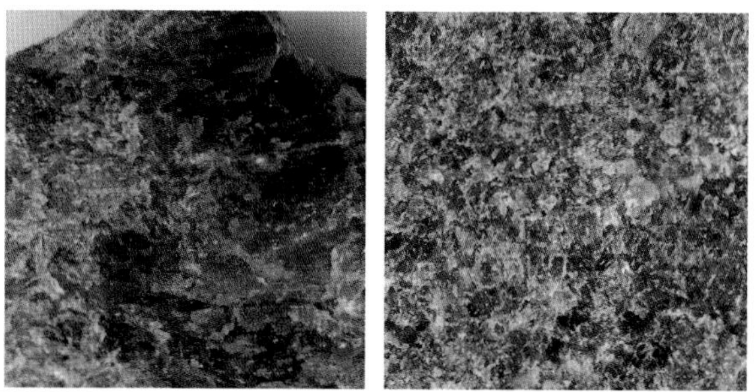

continued overleaf

Hard monomineralic macrocrystalline rocks

g *continued ...*

The "small-step" cleavage is oblique (between 30° and 60°), amphibolic minerals, mostly hornblende: → **Hornblendite** (see also 23b)

Polymineralic-porphyritic rocks with a macro- or microcrystalline matrix

19 a–g

from 5e and 14c

a The fine-grained matrix is mostly light beige to rust red, and the hard, disseminated minerals can be quartz (rounded, gray to brown, transparent) or alkali feldspar (small whitish to reddish blocks), to which a few dark minerals such as biotite and hornblende can be added: → **Rhyolite** (alternate names: **quartz porphyry** or **quartz-feldspar-porphyry**) (see also 4e, 13c)

Round crystallization spherulites can also be present (varioles), mm to cm in size, structured either concentrically or according to a radiating pattern.

19

b The fine-grained matrix is light and hard, gray to brownish; the disseminated mineral is alkali feldspar (small white to reddish blocks), and additionally a few dark minerals like biotite or hornblende can be present:

→ **Trachyte** (see also 4d, 13c) ⊖

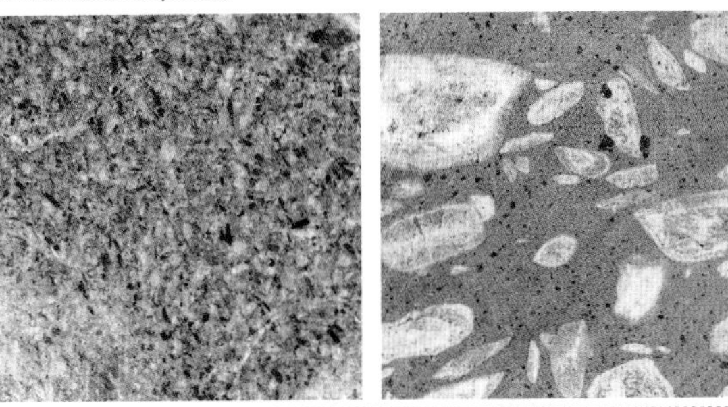

c The fine-grained matrix is gray, greenish gray, or blackish gray; the disseminated minerals are whitish to greenish plagioclase and black hornblende:

→ **Andesite** (see also 4d, 13a) ⊖

Occasionally amygdales (former bubbles, now filled with minerals) are also present.

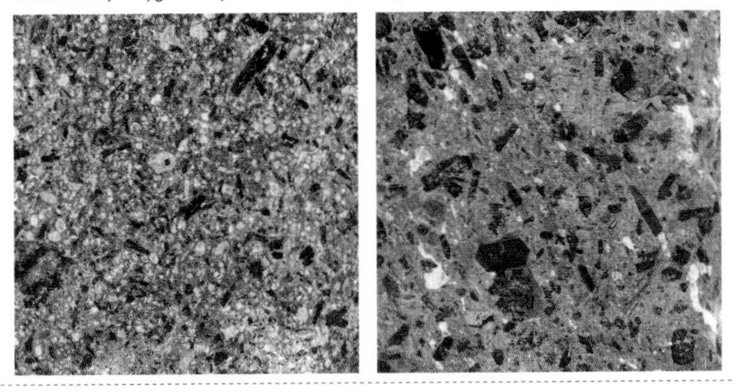

d The fine-grained matrix is dark gray to black; the disseminated minerals are black augite or green olivine, more rarely white plagioclase:

→ **Basalt** (see also 4d, 13a) ⊖

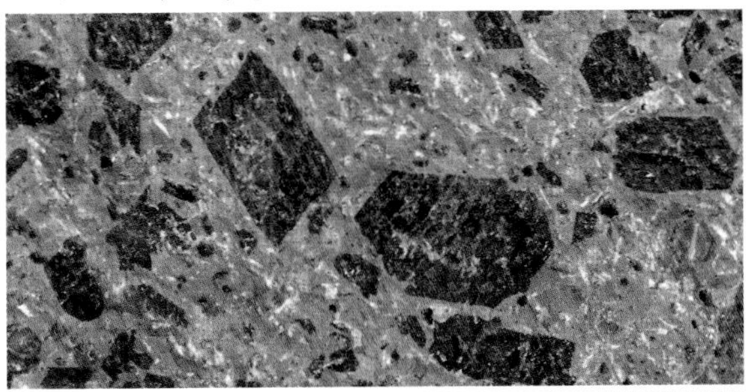

continued overleaf

d *continued ...* ⊖

If light-green olivine is present: → **Olivine basalt** (see also 4d, 13a)

Occasionally amygdales (former bubbles, now filled with minerals) are also present. Round crystallization spherulites can also be present (varioles), mm to cm in size, structured either concentrically or according to a radiating pattern.

e The fine-grained matrix is dark brown to dark blackish green; disseminated minerals are often black hornblende and/or dark brown biotite: → **Lamprophyre** always present as a dike rock! (see also 13b) ⊕ ⊖

Lamprophyre dikes are peculiar, but they are quite common in plutonic complexes. They differ from basalt/andesite primarily in their high potassium content. A whole series of lamprophyre types (minette, vogesite, kersantite, spessartite, etc.) can be distinguished.

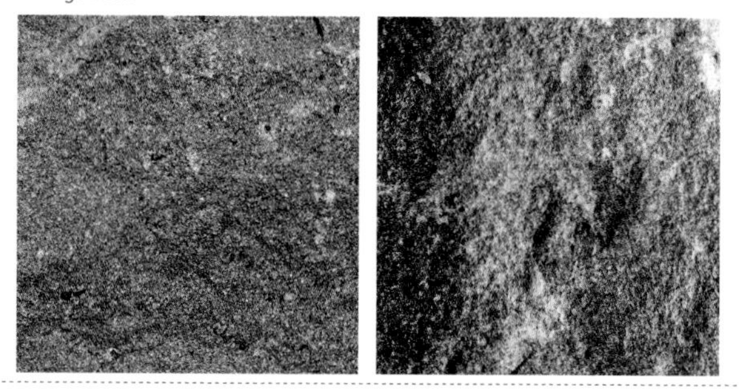

19

f The fine-grained matrix exhibits shades of green, from light to blackish green, often also spotty or streaky; can be scratched without difficulty; disseminated minerals with luster on the cleavage planes, often also black ore minerals:

→ **Serpentinite** with pyroxene phenocrysts (diallage) (see also 11e, 12d, 42a) ⊖

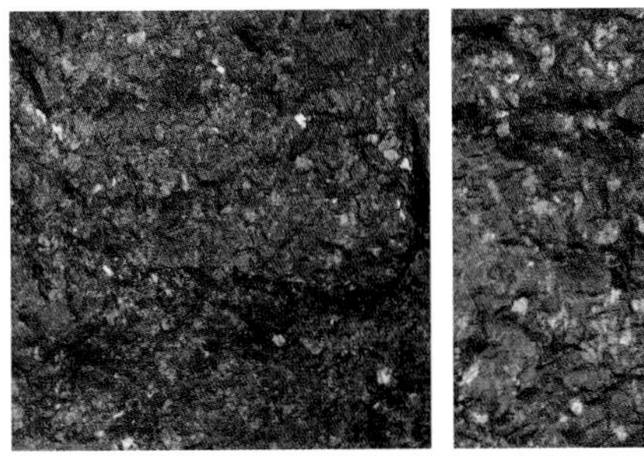

g Large brownish-gray to copper-like crystals with significant luster and good fissility are embedded in a very fine-grained, milky murky, whitish to light green, hard matrix:

→ **Diallage-gabbro** (see also 27c) ⊖

The large crystals are slightly hydrothermally modified igneous pyroxenes (diallage); the milky matrix has been fully transformed into a microcrystalline web of saussuritized plagioclase.

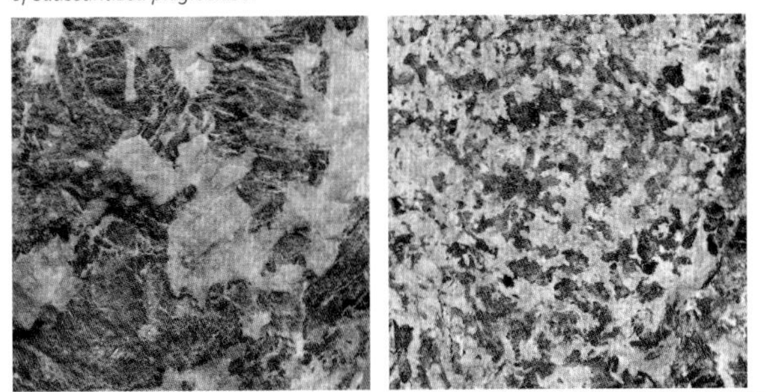

Polymineralic macrocrystalline rocks

20 a–b

Ch. / page

from 14b

a The rock contains minerals that are recognizable with the naked eye or a magnifying glass; it can be found with or without cleavage.

Examples:
granite (left)
garnet–mica schist (right)

⊕ ⊖ → 22 / 114

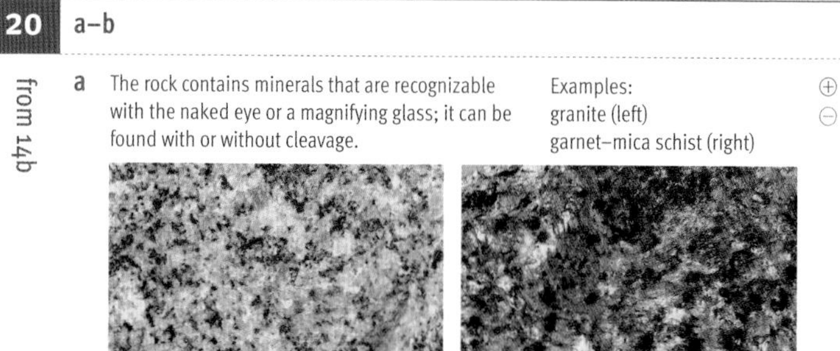

b The rock is grainy massive, often with a sandstone-like appearance. It can exhibit coarse to fine stratification and contains components (rock or mineral pieces) in a matrix that can be only slightly to significantly finer; the components are slightly to well rounded.

Examples:
arkose sandstone (left)
fine-grained graywacke (right)

→ 21 / 113

Arkoses and graywackes

21 a–b

from 20b and 7a

a The rock is relatively light, mostly whitish to reddish, sometimes also really rust red. Feldspar and quartz components are placed in a fine-sandy or argillaceous matrix: → **Arkose = feldspar sandstone** ⊖

Certain arkoses can easily be confused with rhyolites (see 13c, 19a). Very "immature" arkoses from weathered granites (very short transport route) can still almost resemble granites.

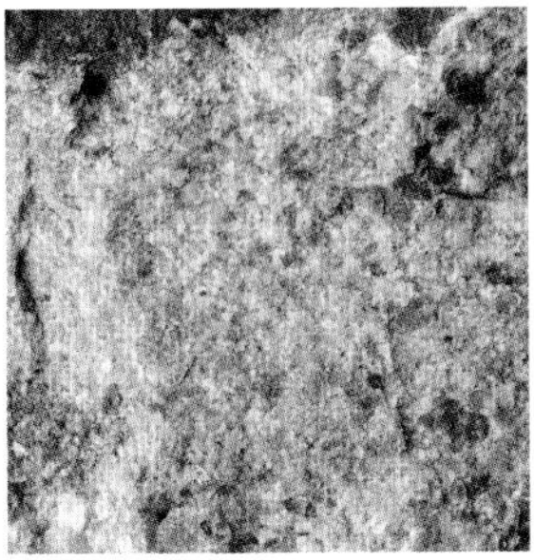

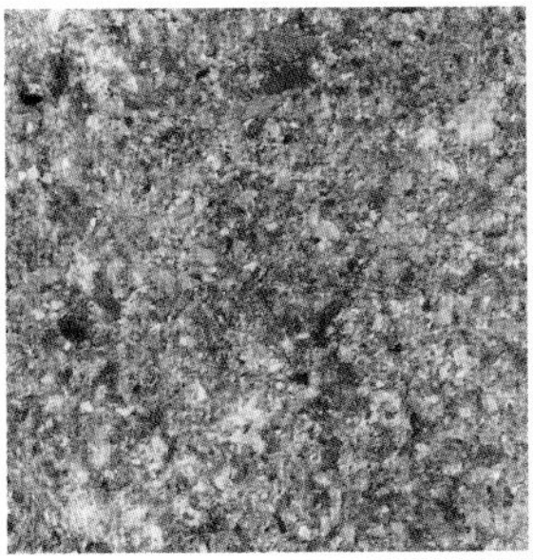

b The rock is greenish gray, gray to blackish; often contains quartz grains and a greater or lesser amount of feldspar, as well as dark minerals such as hornblende, pyroxenes, or biotite in a fairly fine, darker matrix: → **Graywacke/lithic sandstone/glauconitic sandstone** ⊕ ⊖

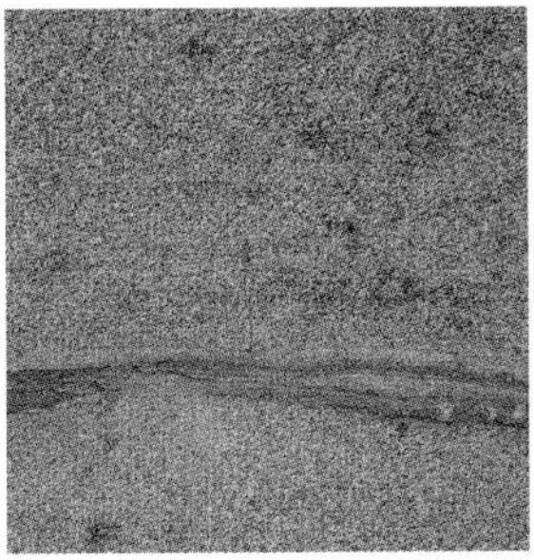

Polymineralic macrocrystalline rocks, igneous or metamorphic

22 a–h

Ch. / page

from 20a

a The rock is completely crystalline, massive directionless (without schistosity, but can also be banded):

→ **Plutonite** (more common case, left photo)

→ **Metamorphic fels** (rarer case, right photo

→ $\frac{23}{118}$

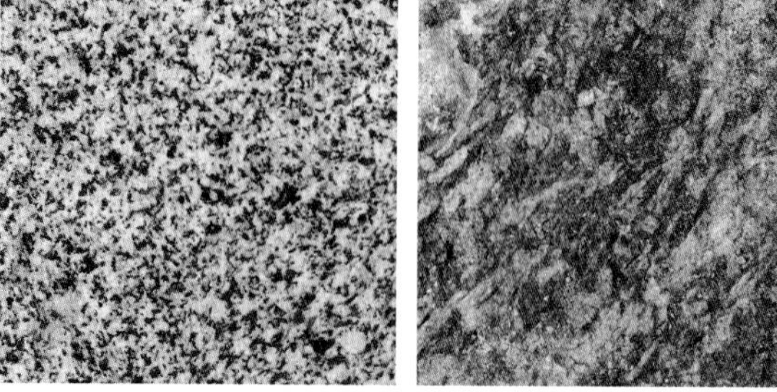

b The rock exhibits more or less rounded quartz grains, sometimes also other mineral grains, white mica among others, in a soft matrix that can be microcrystalline or spathic:

→ **Quartz sandstone/ gritstone** (left) with calcite cement (see also 18a) **Calcareous sandstone** (right) (a wide spectrum of mixtures of lime, sand, and clay is possible!)

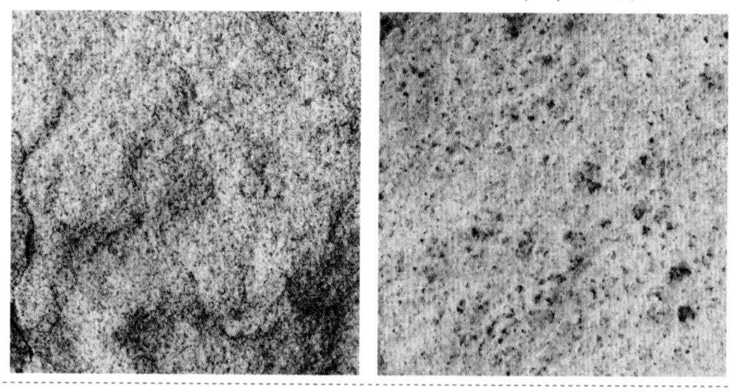

22

Ch. / page

c The rock is fine to medium grained with close-standing schistosity (foliation planes are flat, gaps less than 1 cm); it contains a large proportion of sheet silicates: → **Schistose rocks/slates** → 29 / 136

Schists/slates can also exhibit fine corrugations (crenulations). Shales of somewhat higher metamorphic grade can also exhibit metamorphically grown crystals (blasts); in the case of garnets these appear almost like warts on the foliation planes

d The rock is fine to coarse grained, with either a wavy schlieren-like or a strictly parallel foliation because of aligned, macroscopically visible mica lamellae; the percentage of mica is below ca. 20%: → **Gneisses** → 30 / 142

The rocks are often irregularly banded or show "augen" structures; in outcrops they are present as more or less thick slabs, which were frequently used for wall and house construction in earlier times; the rocks can also exhibit bands of different materials.

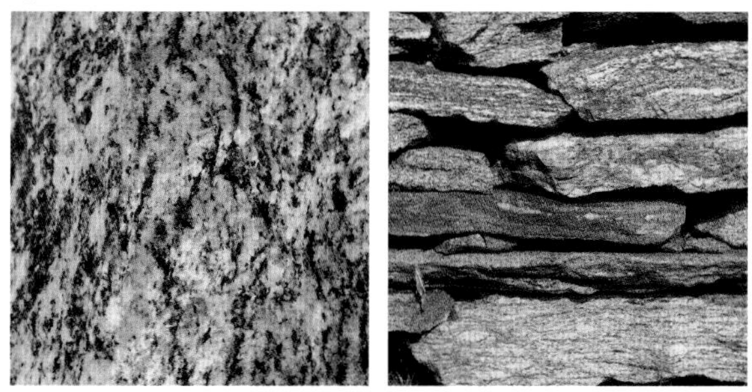

continued overleaf

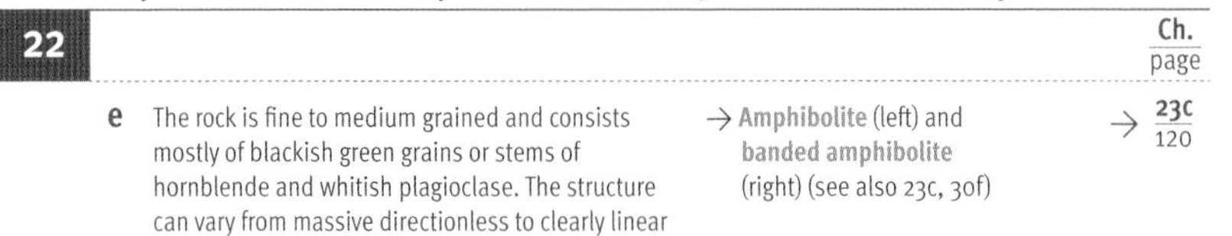

Polymineralic macrocrystalline rocks, igneous or metamorphic

22

Ch. / page

e The rock is fine to medium grained and consists mostly of blackish green grains or stems of hornblende and whitish plagioclase. The structure can vary from massive directionless to clearly linear (the orientation of the hornblende blades), foliated (when biotite is present), or banded in alternating light-dark layers:

→ **Amphibolite** (left) and **banded amphibolite** (right) (see also 23c, 30f)

→ **23c** / 120

f The rock is white or very light, massive directionless (nonstratified, nonfoliated), coarse grained (grains often over 1 cm); consists of feldspar and/or quartz and/or calcite, sometimes with additional minerals:

→ **Pegmatite** or ⊖

→ **28** / 134

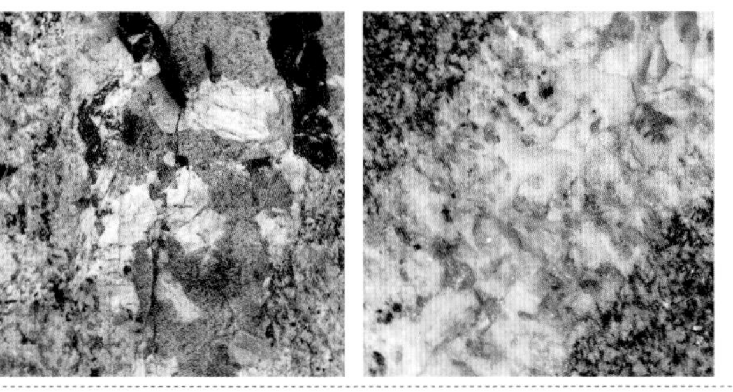

→ **Hydrothermal veins**

→ **43** / 156

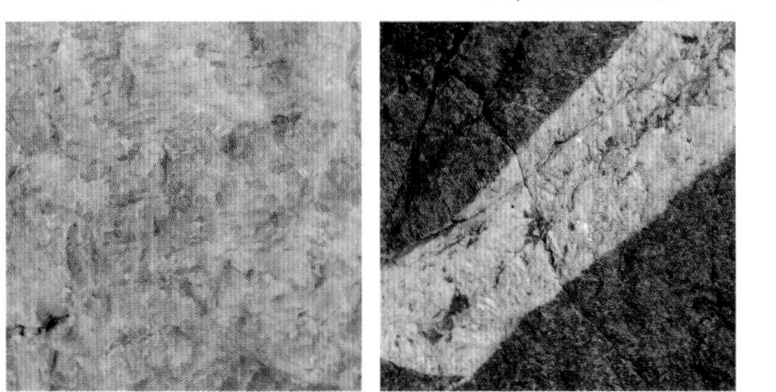

22 Ch./page

g The rock is medium to coarse grained; on the outcrop scale consists partly of components with a metamorphic foliation and/or banding and partly of components with a massive-directionless texture (igneous texture): → **Migmatite** ⊖ → 31/146

Migmatites are high-grade metamorphic rocks in which local melting has occurred because of high temperatures. They represent a transition between metamorphic and plutonic rocks

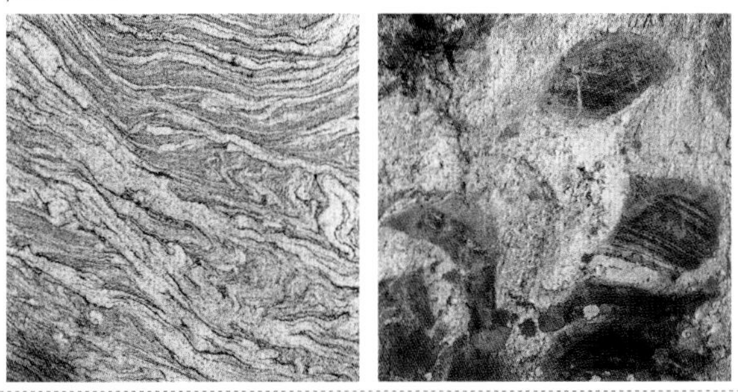

h The rock consists overwhelmingly of ore minerals. These rocks are normally of a high specific weight, and the ore minerals give themselves away by their color and shimmer. In between the ore minerals there can be more or less pronounced amounts of non-ore minerals (often quartz, calcite, or feldspar): → **Ore** ⊕ ⊖ → 40/148

continued overleaf

Macrocrystalline plutonites and metamorphites

23 a–j | Ch./page

from 22a

a The rock contains the light-colored minerals quartz and/or feldspars as well as differing amounts of dark minerals such as biotite, muscovite, hornblende, pyroxenes, and olivine: → **"Normal" plutonites** ⊖ → 24/127

The amount of dark minerals can be as high as **90%**, *but these very dark rocks are still classified within this group!*

b The rock consists mostly (> 90%) of dark minerals such as hornblende, pyroxenes, and olivine; the rocks have a higher specific weight than "normal" rocks (around 3.3 g/cm³): ultramafic plutonites.

The principal mineral is olivine: → **Peridotite** (see also 18d) ⊖

The entire upper mantle consists of peridotite rock.

Such ultramafic rocks can also be formed under very high metamorphic conditions **(ultramafic fels)**. *Differentiation is usually possible only in the geological context. In the Alps the peridotites of the area around Ivrea (Finero, Balmuccia, Baldissero) are well known, and they partially exhibit metamorphic structures and can contain considerable amounts of the light brown Mg-biotite phlogopite.*

23

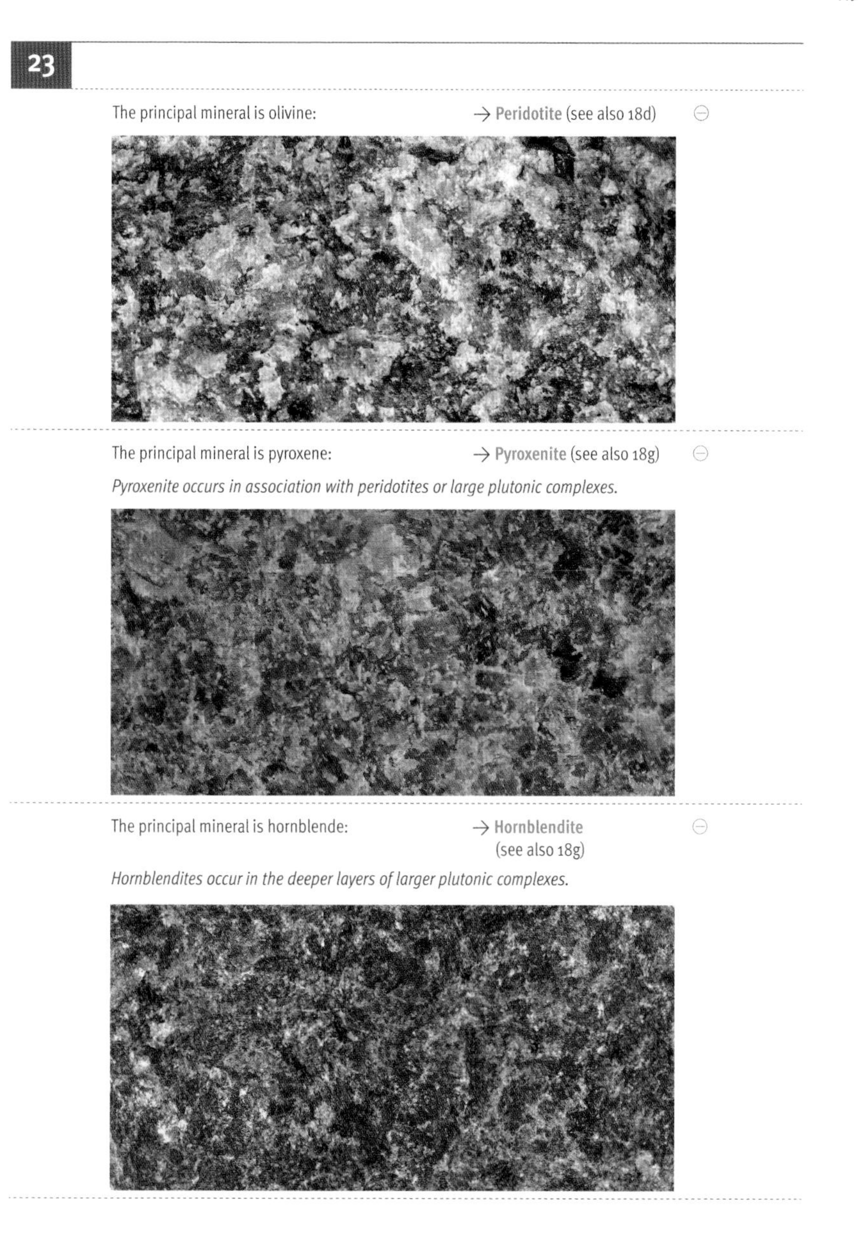

The principal mineral is olivine: → **Peridotite** (see also 18d) ⊖

The principal mineral is pyroxene: → **Pyroxenite** (see also 18g) ⊖

Pyroxenite occurs in association with peridotites or large plutonic complexes.

The principal mineral is hornblende: → **Hornblendite** (see also 18g) ⊖

Hornblendites occur in the deeper layers of larger plutonic complexes.

continued overleaf

Macrocrystalline plutonites and metamorphites

23

from 22e

c The rock is fine to medium grained and consists of whitish to greenish feldspar (plagioclase) and black (to greenish-black) hornblende; it can also contain biotite or garnet. It exhibits a more or less pronounced planar or linear texture because of aligned blades of hornblende and/or banding with lighter and darker layers: amphibolite.

→ **Amphibolite** (= a high-grade metamorphic basaltic rock) (see also 22e, 30f) ⊖

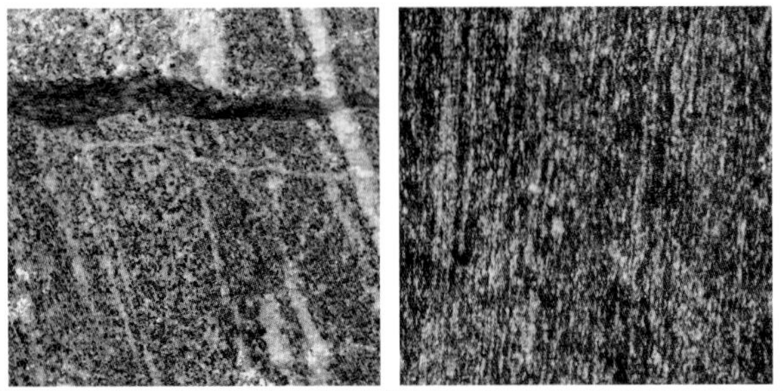

→ **Banded amphibolite** (a widespread variant of amphibolites; the amphibolite layers alternate with the lighter feldspar layers) ⊖

→ **Garnet amphibolite,** in cases where garnet is clearly present ⊖

23

→ **Epidote amphibolite,** in cases where epidote is clearly present ⊖

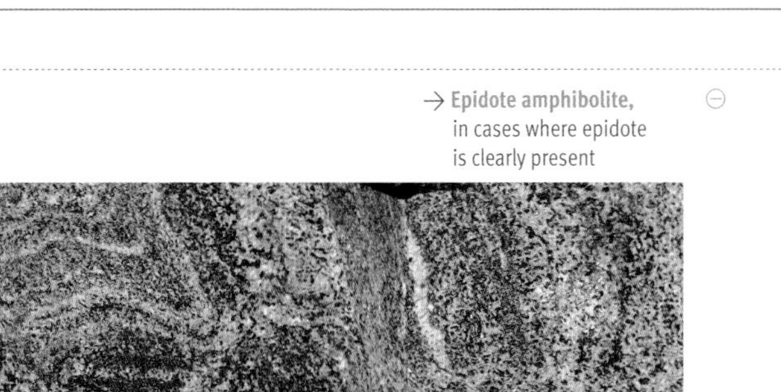

d The rock is fine to medium grained and consists especially of green minerals such as chlorite, actinolite, and epidote, together with whitish albite-feldspar. It frequently contains small fissures and veins with the same minerals:

→ **Greenstone** (greenschists if slightly schistose); weakly to moderately metamorphic basalt (see also 12c, 29e) ⊕ ⊖

*Metabasalts that formed in hydrothermal metamorphic processes at midocean ridges and were transformed to greenstones were formerly called → **spilites**. Greenschists can exhibit all transitional forms from massive compact to strongly foliated (see 12c and 29e).*

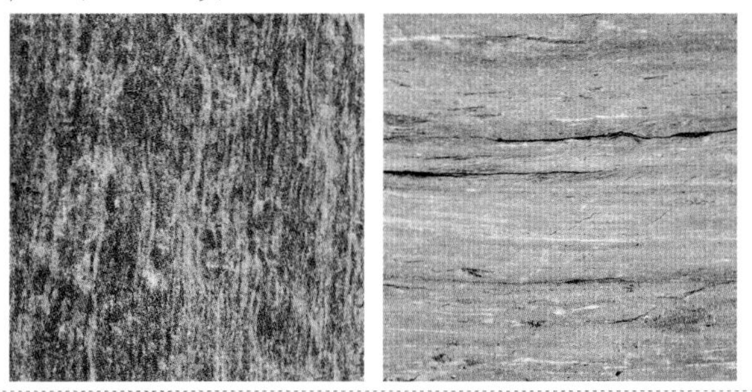

continued overleaf

e The rock is fine to medium grained and consists of roughly equal amounts of red garnet and green yroxenes (omphacite). Additionally blackish-purple glaucophane, silvery white mica, white zoisite prisms, or black chloritoid lamellae may also be present in smaller quantities. A relatively heavy rock:

→ **Eclogite** (high-pressure metamorphic basaltic rock). Eclogites can also exhibit a banding of materials, either primary or because of metamorphic processes. ⊖

f The rock consists of a grainy groundmass of greenish olivine (it can be confused with quartz; the greenish color is decisive!), in which rounded red garnets are enclosed. Light green pyroxenes may also be present. The garnets may exhibit dark retrograde alteration borders.

→ **Garnet peridotite** (= high-pressure metamorphic peridotite) ⊖

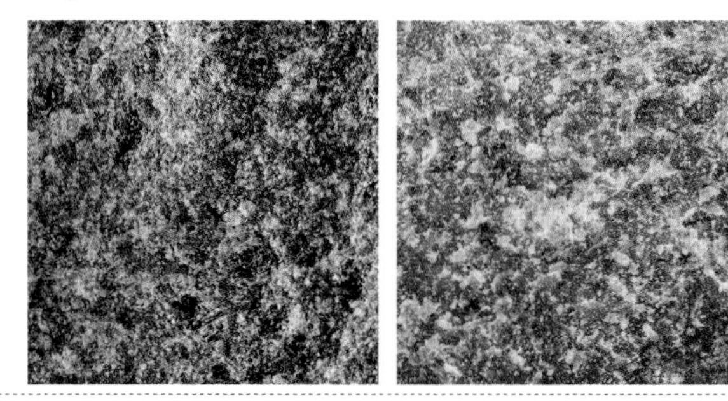

23

g The rock is grainy with clearly recognizable minerals, often layered or banded, but also directionless. Contains white feldspar, often quartz; this is accompanied by greatly varying amounts of dark minerals, mostly garnet and pyroxenes, yet other minerals can also be present (for instance kyanite, sillimanite, cordierite, rutile, and ilmenite; but never muscovite!): granulite

You can distinguish between: → **Felsic granulites** ⊖

Light-colored minerals (feldspar/quartz) with less than 30% dark minerals:

Granulites are high-grade metamorphic rocks named after their granulite facies (temperatures above 700o C). Demarcation with plutonic rocks is not always easy.

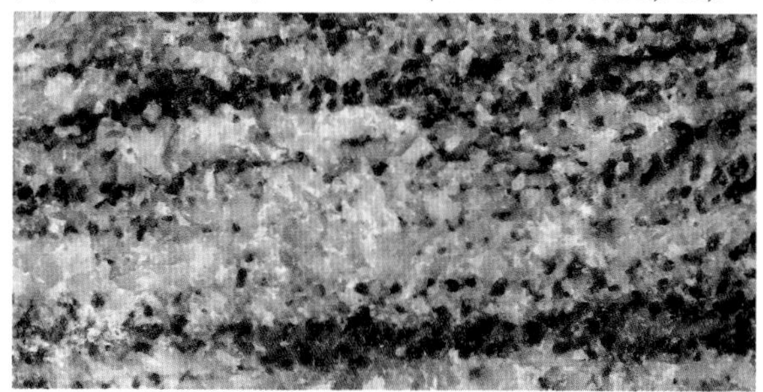

continued overleaf

Macrocrystalline plutonites and metamorphites

Darker rocks with over 30% darker minerals: → **Mafic granulites** ⊖

Mineral names are ordered by ascending prevalence, for instance felsic orthopyroxene garnet granulite.

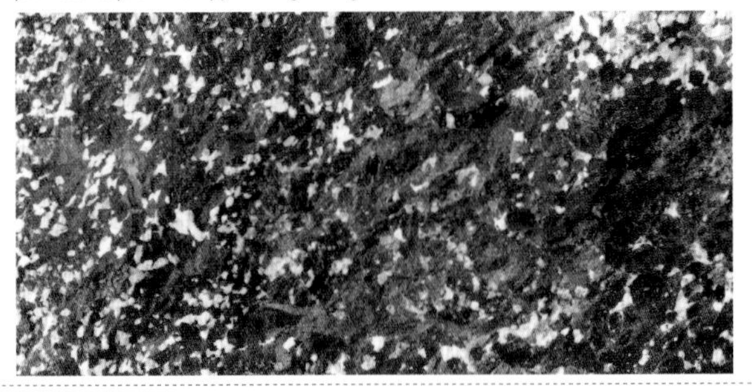

h The rock contains larger amounts of other minerals such as garnet, cordierite, calcite, vesuvianite, diopside, epidote, glaucophane, and sapphirine, sometimes with quartz and/or feldspars: → **High-grade metamorphic fels** left: granulite facies metapelite (cordierite potassium feldspar fels) right: sapphirine orthopyroxene garnet fels ⊕ ⊖

Mineral names are ordered by ascending prevalence, for instance cordierite garnet quartz fels.

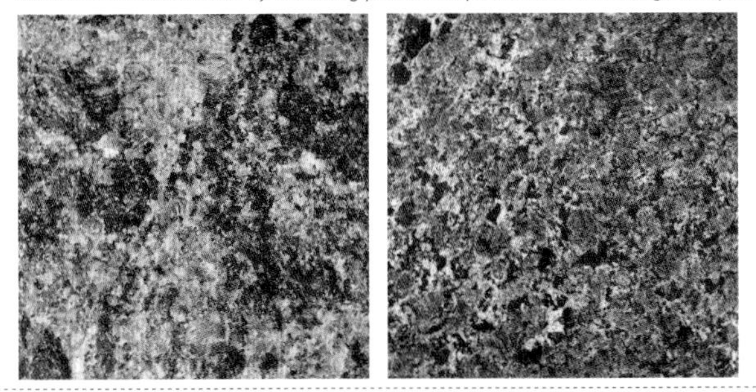

i **Special cases:**
The rock contains many calcium-rich silicate minerals like garnet, epidote, amphibole, pyroxene, and vesuvianite; sometimes also calcite and chlorite: → **Calc-silicate fels:** high-grade metamorphic marly rock with many Ca-minerals (epidote, vesuvianite, garnet, and calcite). ⊕

23

→ **Skarn:** transformation product of calcareous rocks in the presence of hot plutonic fluids.

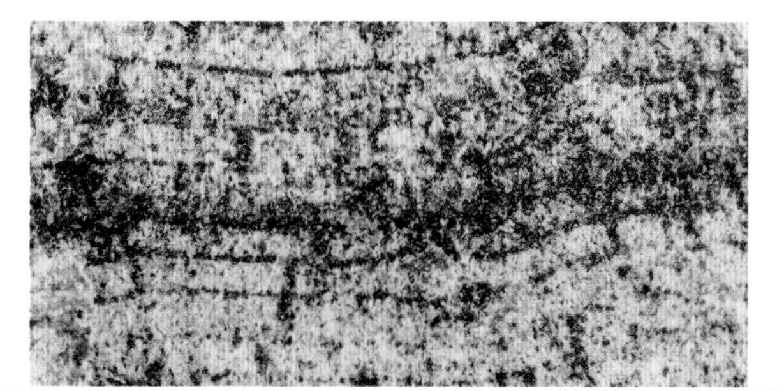

→ **Rodingite:** fine-grained dense transformation product of basalt dikes in serpentinites, often with coarse-grained colored areas with a lot of light green epidote and reddish-brown garnet, brown vesuvianite, and dark green chlorite

continued overleaf

23

j The rock is spotty, exhibiting white, yellowish greenish, and green. Often heterogeneous, flaser-like, chaotic; white parts are often talc (very soft, smudgy):

→ **Impure soapstone (Gilt-/Lavez-/Ofenstein):** mixed rocks of talc/chlorite/serpentinite/amphibole, partly with carbonate. Was used in the construction of ovens. ⊕ ⊖

Impure soapstone is a reaction product of serpentinite with siliceous surrounding rock. The name "Ofenstein" (oven rock) comes from its widespread use in wood-burning stoves to store warmth

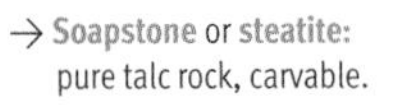

→ **Soapstone** or **steatite:** pure talc rock, carvable. ⊖

Plutonites (without ultramafics)

24 a–c Ch./page

from 23a

a The rock contains significant amounts of quartz (at least ca. 20%): → **Family of the granitoid rocks** ⊖ → 25/128

b The rock contains less than ca. 20% or no quartz → **Plutonites that contain either little or no quartz** ⊖ → 27/132

c The rock contains no quartz and no feldspars but feldspar substitutes (foids, often nepheline), a very rare special case; foids can be identified only by specialists. → **Foid plutonites** (for further classification see the QAPF diagram on p. 175) ⊖

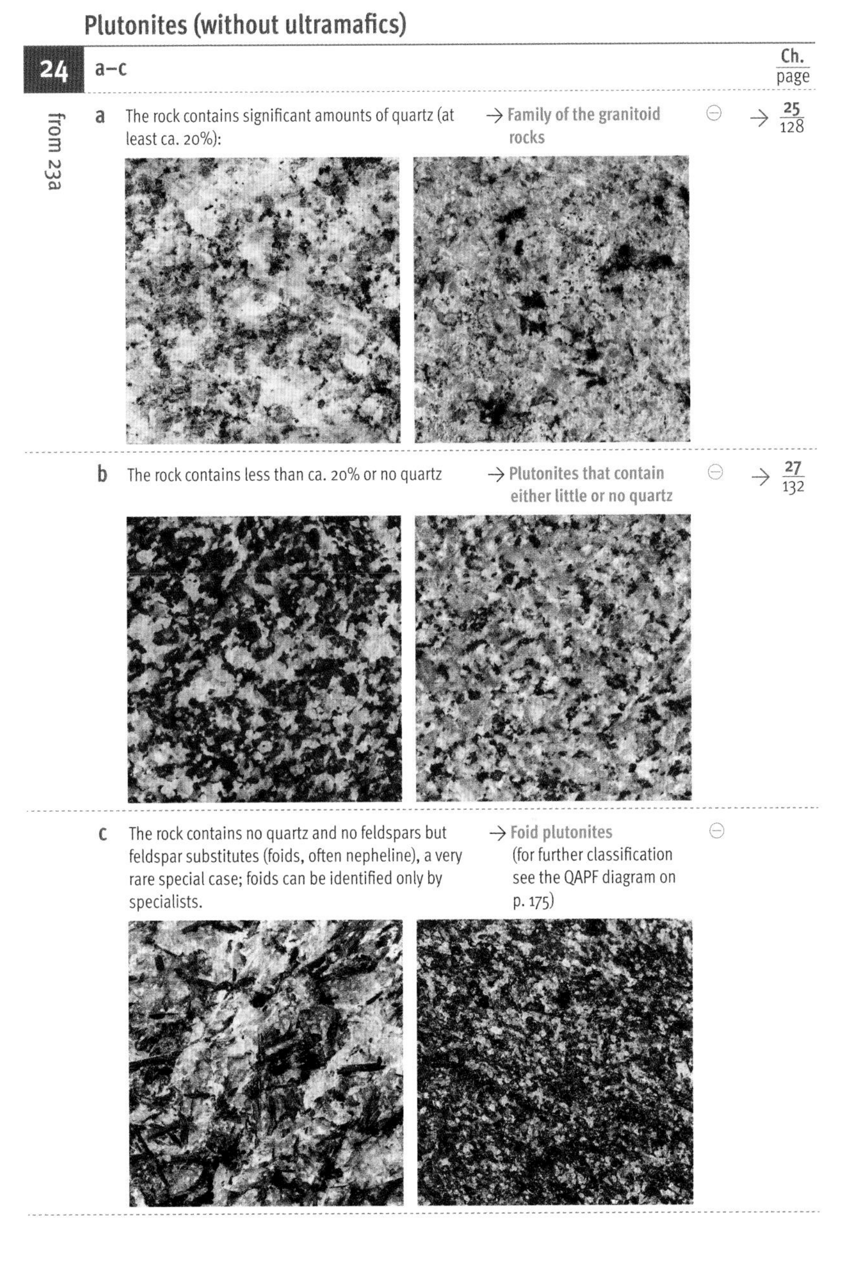

Granitoid rocks in general

25 a–c

from 24a

Family of **granitoid rocks**. For further identification you need to understand the difference between potassium feldspar and plagioclase. Under favorable circumstances these can be differentiated macroscopically (but certainly not always).

Potassium feldspar
- tends to exhibit larger grains, sometimes actually phenocrysts in the shape of small blocks up to several cm in size (→ porphyritic texture)
- often exhibits a suture of the twin plane midgrain (a separate gleam for the two grain halves)
- tends to be reddish to rust brown

Plagioclase
- tends to be greenish to yellowish
- the gleaming cleavage surfaces are mostly recognizable

There are also granitoid rocks that contain only one of the two feldspars!

a The rock is fine grained to very fine grained, light to white, does not bear any micaceous minerals, under certain circumstances can bear some garnet: → **Aplite (dike rock)** ⊖

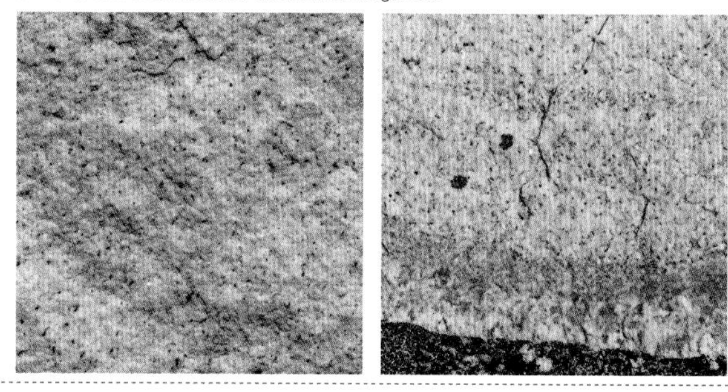

b Differentiation between the two feldspars is not possible: → Rock in the **granite family ("granitoids")** ⊖

Further classification according to the amount of dark minerals contained, such as dark or white micas (biotite, muscovite) or hornblende:

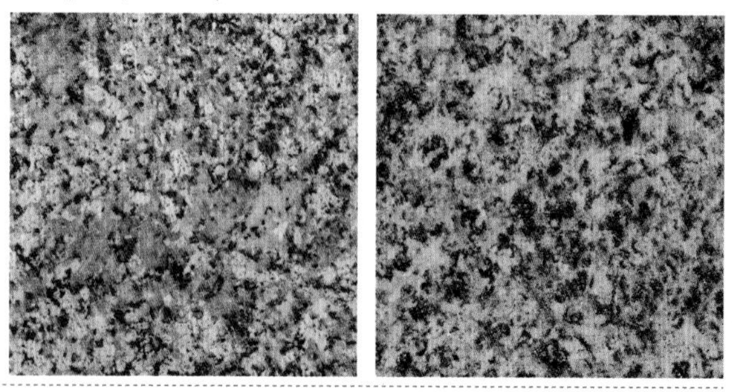

Up to 30% dark minerals by volume: → Additional adjective **leucocratic** (left photo)

30–60% dark minerals by volume: → Additional adjective **mesocratic** (right photo)

Over 60% dark minerals by volume: → Additional adjective **melanocratic**

Example: mesocratic hornblende biotite granitoid:

If the potassium feldspars are present in the form of significantly larger, more or less well-crystallized (idiomorphic) crystals, then it is a porphyritic texture: → **Porphyritic granitoids** ⊖

C Differentiation of the two feldspars is possible. → **Granitoid rocks in detail** ⊖ → 26/130

Left: Baveno granite; potassium feldspars are reddish, plagioclase whitish greenish.

Right: Mont Blanc granite; potassium feldspars are whitish, plagioclase grayish green.

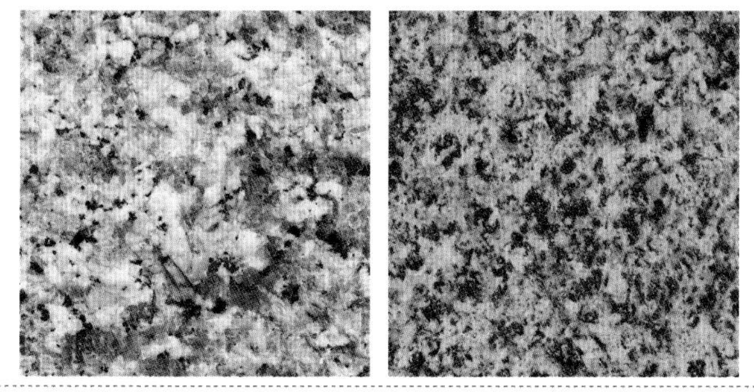

Granitoid rocks in detail

26 a–e

Additional prefixes can also be used to point to increasing amounts of dark minerals present.

Example: *mesocratic hornblende biotite granodiorite*

from 25c

a Feldspars: amount of potassium feldspar 100–90%, plagioclase 0–10%: → **Alkali feldspar granite** ⊖

Further classification into leuco-, meso-, and melanocratic types or with mineral prefixes as in 25b.

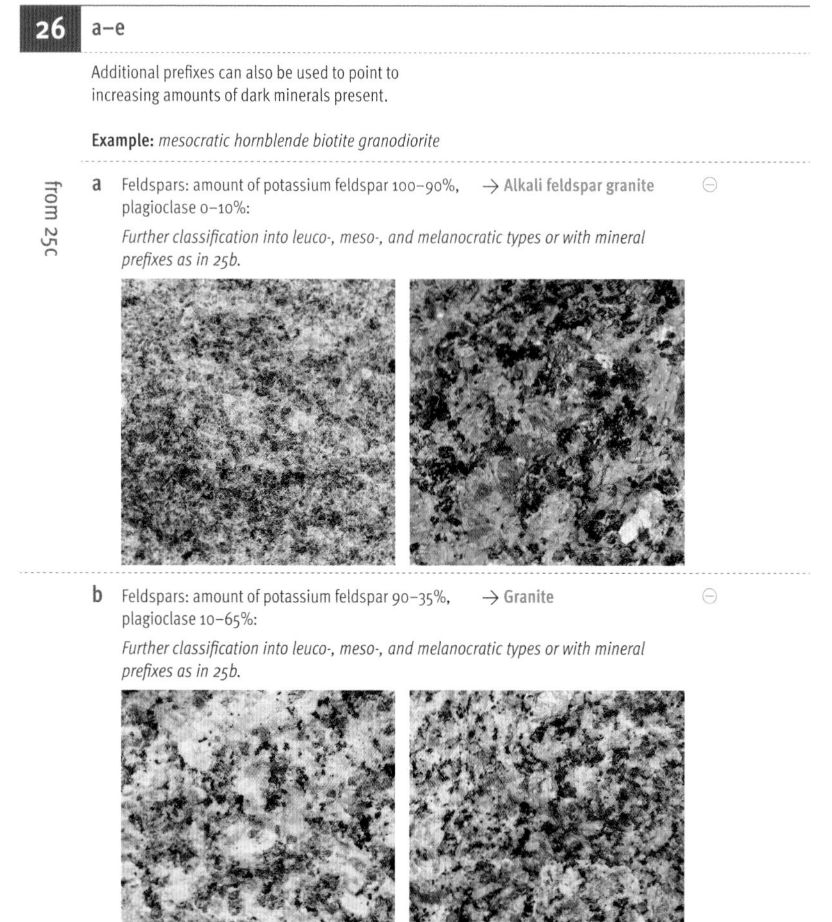

b Feldspars: amount of potassium feldspar 90–35%, plagioclase 10–65%: → **Granite** ⊖

Further classification into leuco-, meso-, and melanocratic types or with mineral prefixes as in 25b.

c Feldspars: amount of potassium feldspar 35–10%, plagioclase 65–90%: → **Granodiorite** ⊖

Further classification into leuco-, meso-, and melanocratic types or with mineral prefixes as in 25b.

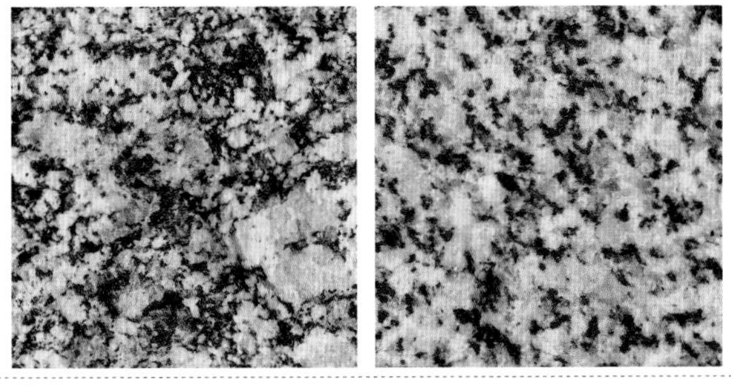

d Feldspars: amount of potassium feldspar 10–0%, plagioclase 90–100%: → **Tonalite** ⊖

Further classification into leuco-, meso-, and melanocratic types or with mineral prefixes as in 25b.

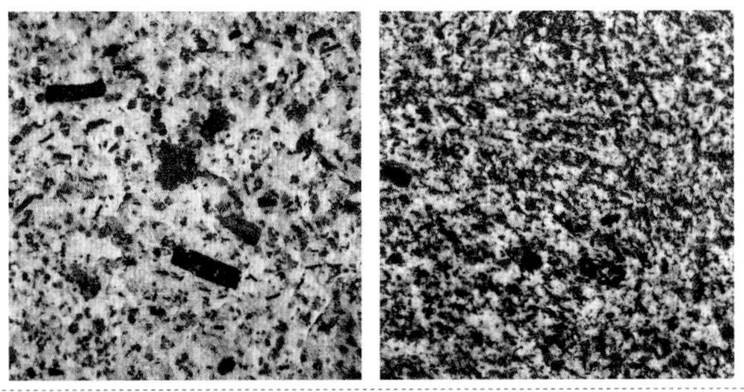

e If the potassium feldspars are present in the form of significantly larger, more or less well-crystallized (idiomorphic) crystals, then it is a porphyritic texture: → **Porphyritic rocks** ⊖

Plutonites with little to no quartz

27 a–d

from 24b

For the differentiation of feldspars see the introduction to 25.
When it is not possible to differentiate on the basis of a hand specimen, the geological context should be used as well, and from these elements conclusions can be drawn as to the main feldspar.

a Feldspars: the amount of potassium, feldspar is 100–65%, plagioclase 0–35%: → **Syenite** (left) or **quartz-syenite** (right) (if quartz › 5%) ⊖

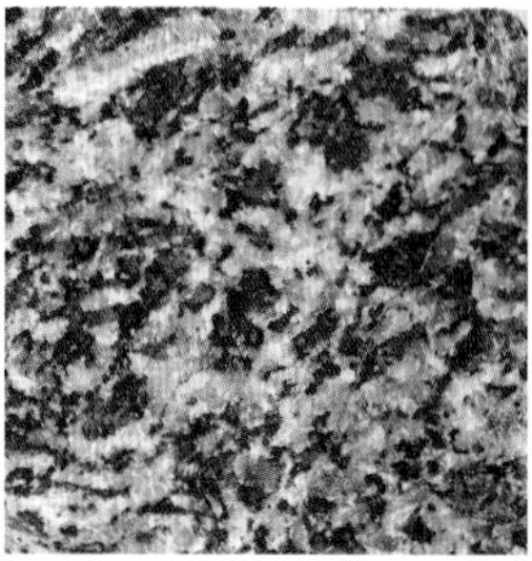

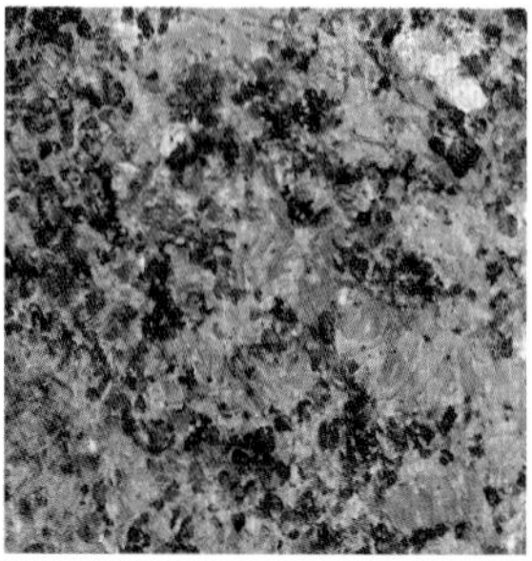

b Feldspars: the amount of potassium feldspar is 65–35%, plagioclase 35–65%: → **Monzonite** or **quartz monzonite** (if quartz › 5%) ⊖

Further classification into leuco-, meso-, and melanocratic types or with mineral prefixes as in 25b.

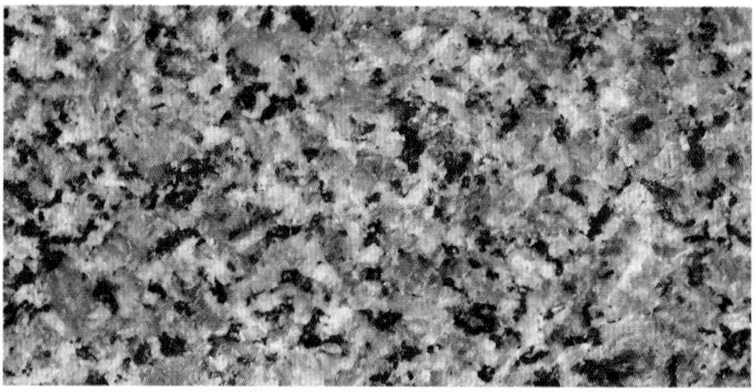

c Feldspars: the amount of potassium feldspar is 35–10%, plagioclase 65–90%: → **Monzodiorite** or **quartz monzodiorite** (if quartz › 5%) ⊖

Further classification into leuco-, meso-, and melanocratic types or with mineral prefixes as in 25b.

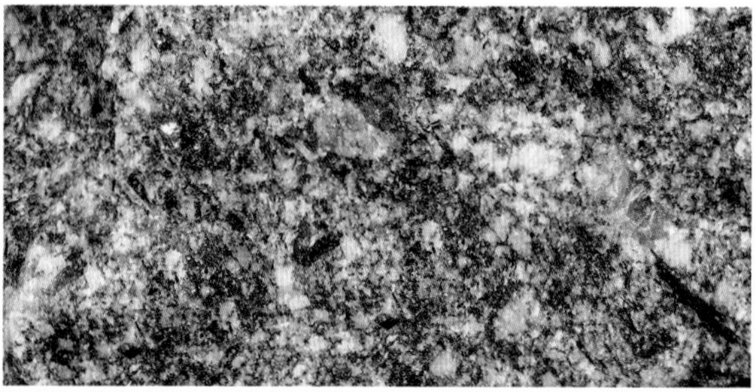

27

d Feldspars: the amount of potassium feldspar is 15–0%, plagioclase 85–100%: **diorite** or **gabbro**.

Differentiation of diorite vs. gabbro:

→ **Diorite** Mostly fine to medium grained; dark minerals hornblende and/or biotite.

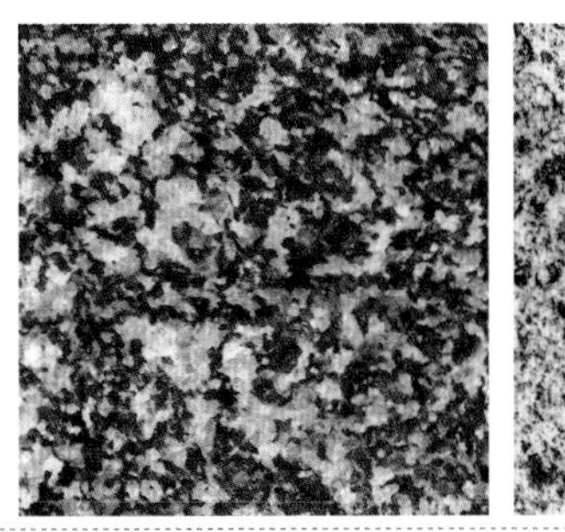

→ **Gabbro** Mostly medium to coarse grained; dark minerals pyroxenes and/or olivine. Pure olivine gabbro = **troctolite** (right), orthopyroxene-rich gabbro = **norite** (left)

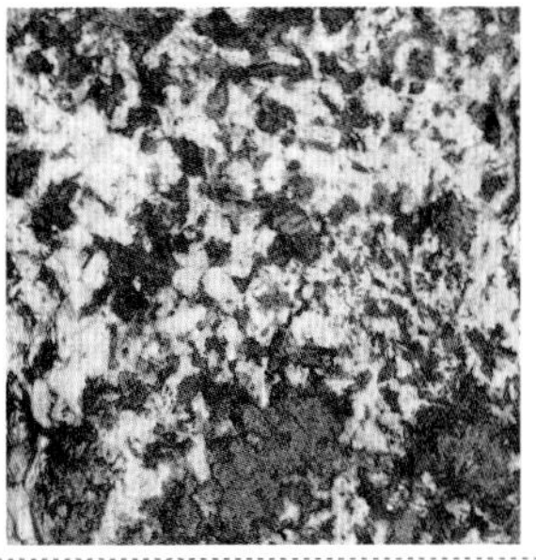

continued overleaf

Plutonites with little to no quartz

Special case: → **Anorthosite** The rock consists almost exclusively of plagioclase; see also 18f. ⊖

Further classification into leuco-, meso-, and melanocratic types or with mineral prefixes as in 25b.

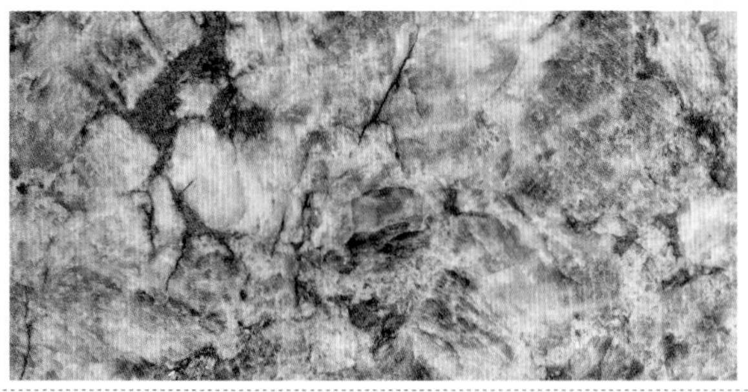

Light-colored, coarse-grained, polymineralic rocks

28 a–b

from 22f

a The rock is white/light gray to pink and consists of gray quartz and white feldspar crystals closely intergrown with one another: → **Pegmatite (-dike)** ⊖

Light-colored, coarse-grained, polymineralic rocks

28

Ch. / page

a *continued ...* ⊖

The intergrowth patterns can exhibit rune-like geometries. The crystal grains are large to very large and can reach several cm. The dike structure in the plutonic, usually granitic mother rock can be recognized in the outcrop.

Frequently with large crystal sheets of muscovite or biotite. Pegmatites can be of either plutonic or high-grade metamorphic origin. Plutonic pegmatites are formed from watery residual melts with many trace elements and can therefore also contain rarer minerals like tourmaline (black schorl), garnet, aquamarine, or spodumene, also uranium minerals.

b The rock consists of different amounts of coarsely crystalline whitish to grayish quartz, white feldspar, and/or white to ocher calcite. The rock's context often indicates that the formations represent joint and vein fillings, or concretions, in metamorphic rocks: → **Hydrothermal quartz / feldspar / calcite veins** ⊕ ⊖ → 43 / 156

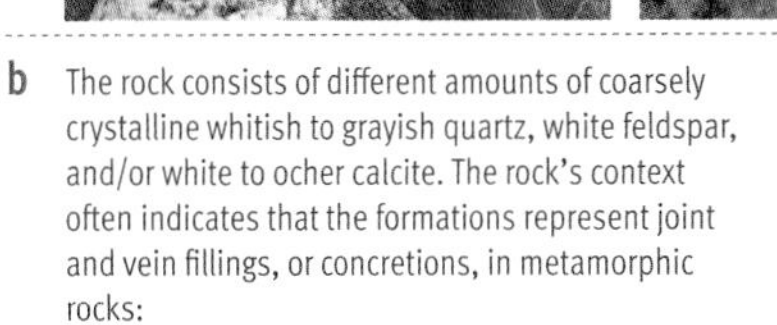

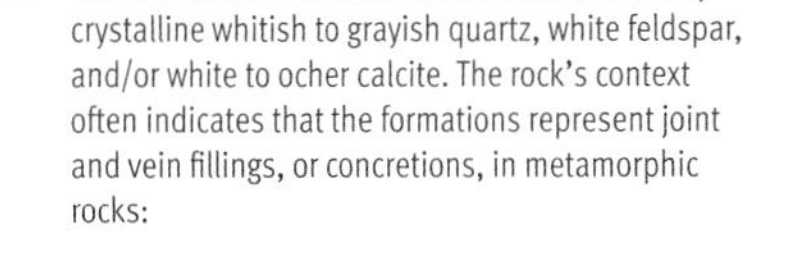

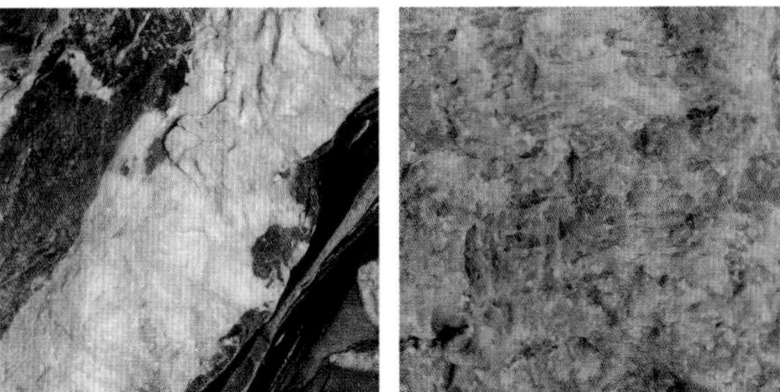

Shales, slates, and schists

29 a–g

from 22c

Shale (almost nonmetamorphic), slate, or schist is a textural concept that proves useful in the field to group microcrystalline to fine-grained metamorphic rocks with a very marked and strictly parallel foliation. Many different rocks (protoliths) can be transformed into slates or schists at low to medium metamorphic grades. A selection of the most common is presented here. If you cannot find what you are looking for, you can simply list the attributes that seem important, as well as the recognizable minerals, and add them before the term "slate" or "schist" to obtain a serviceable descriptive field designation. **Examples:** fine-grained, gray, garnet-bearing two-mica schist; chalky, greenish, sericite quartz slate.

a Mostly gray to black, very fine grained and fine leaved, splitting into regular slabs. Such weak-grade metamorphic argillaceous shales/slates (metapelites) were once valued as roof slabs, shingles, chalkboards, and so on:

→ **Argillaceous shale or slate** ⊖ (see also 9d, 11a)

29

With increasing limestone content transition to: → **Marly shales or slates** to **argillaceous limestone slates** (see also 9d, 11b) ⊕

b Gray to light silvery, fine-grained and finely foliated rocks, rich in very fine-grained micaceous minerals (50% or more): → **Phyllite** (left) or **sericite slate** (right) ⊖

The micaceous minerals include mostly white mica (muscovite), which in its most fine-grained formation is called "sericite." The rocks are characterized by their silklike luster on the cleavage surfaces. Increasing amounts of chlorite give the rocks an increasing green color. More-targeted designations can be made using structural and mineralogical attributes, for instance folded, pyrite-bearing chlorite sericite slate (lower photo).

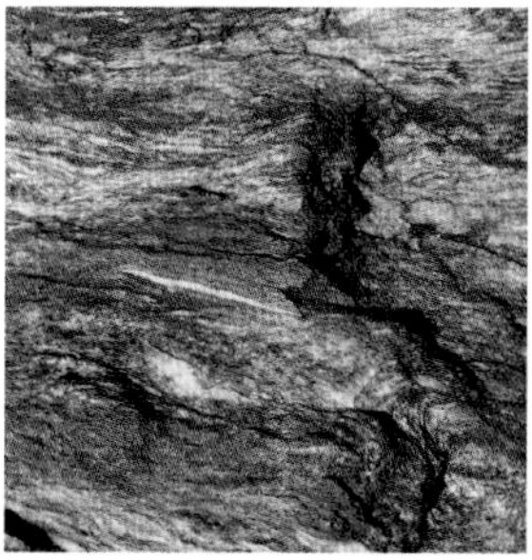

continued overleaf

Shales, slates, and schists

C The higher-grade metamorphic equivalent of the phyllite/sericite slates. The micaceous minerals (white and black mica) aligned on the cleavage surfaces are clearly recognizable:

→ **Mica schist**

The foliation is not as strictly parallel and close standing as with the phyllites. Other minerals, mostly quartz (> 30%), are recognizable between the mica layers.

In higher-grade metamorphic micaceous schists a variety of metamorphic minerals can appear, and they are often present in larger and clearly recognizable crystals ("blasts"), which are sometimes very beautiful and spectacular:

Chloritoid: very shiny black lamellae, harder than biotite

Garnet: reddish-brown crystals (middle and lower photos)

Staurolite: brownish-red stem-like crystals

Kyanite: blue/blue-white lamellar blades

Hornblende: black/blackish-green blades, often sheaf-like

Cordierite: bluish quartz-like grains

More exact designations are based on these additional minerals—for example, staurolite kyanite-bearing garnet two-mica schist.

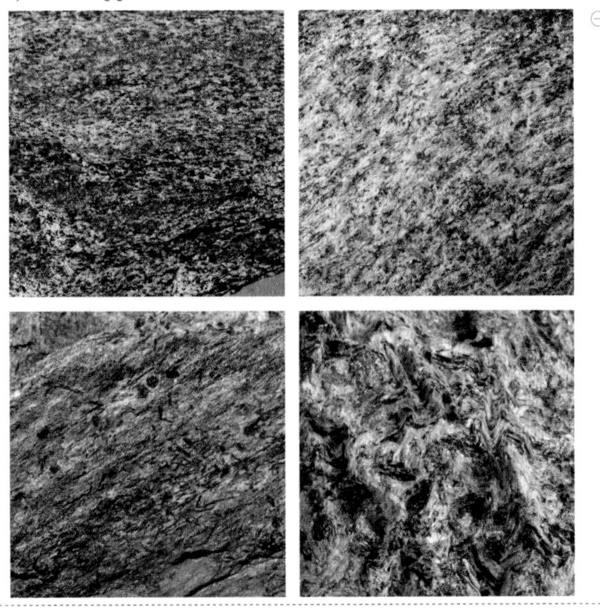

29

C *continued ...* ⊖

Special case 1:
Mica schists that were formed from marly rocks can contain very significant amounts of hornblende. This is sometimes formed into either sheaf-like or broom-like aggregates:

→ **Hornblende schist** ⊕ ⊖

Special case 2:
Light-colored basically quartz-free slate with a whitish fine-grained matrix (albite, zoisite, etc.) and conspicuous light green micas (muscovite that contains chrome = fuchsite):

→ **Fuchsite schist** (= strongly deformed greenschist facies **metagabbro**) ⊖

continued overleaf

Shales, slates, and schists

29

d Schistose-lamellar rocks that contain a lot of calcite marble in finely foliated layers, mostly in strata between the mica-rich layers, in addition to greater or lesser amounts of mica and chlorite: → **Calcareous mica schist (carbonate-silicate schist)** ⊕

All possible kinds of mixtures exist, from almost pure marbles with a low mica content to mica-rich calcite-poor compositions. These, which were originally limestone mudrock sediments, are typical for ocean basin/accretionary wedge deposits at subduction zones and continental collisional settings.

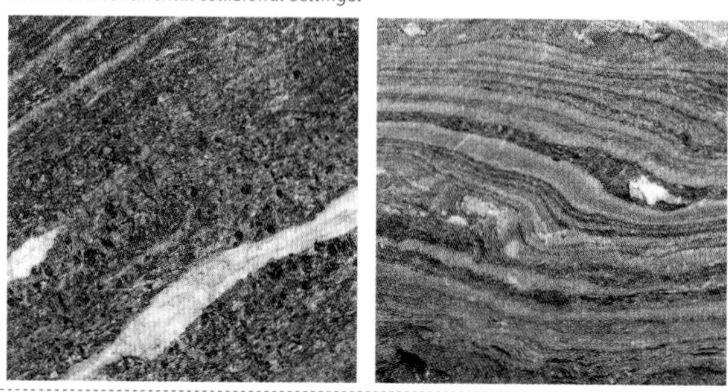

e Greenschists consist of differing amounts of the minerals actinolite, chlorite, epidote, albite, also white mica or calcite. These are medium-grade metamorphic basalts: → **Greenschists** (see also 23d and 12c) ⊕ ⊖

They can be clearly banded and/or foliated to massive. The bands correspond to different amounts of the minerals mentioned above. They can represent original volcanic layers and sometimes highly deformed pillow lavas.

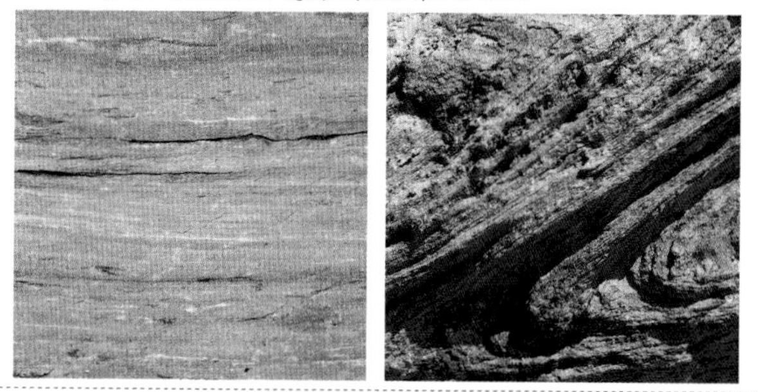

29

f Blueschists are characterized by the blackish-purple to blackish-lilac amphibole glaucophane. They often also contain a pistachio-green epidote; many blueschists are clearly banded, with dark glaucophane and green epidote layers. These are high-pressure metamorphic basaltic rocks:

→ **Blueschists**

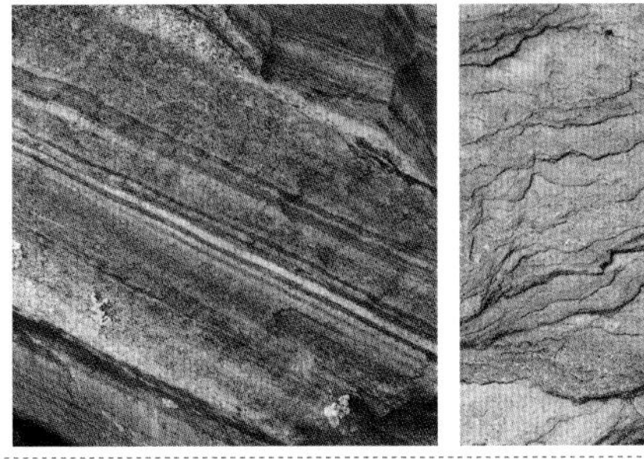

g In shear zones granites can exhibit all the transitions, from granite to granite gneiss to granite schist and even granite mylonite. Granite schists consist of quartz and feldspars, to which finer, newly crystallized muscovite and chlorite (formed from the biotite in the granite) are normally added on the cleavage surfaces. Mylotinized granite is a slabby, fine-grained, greenish rock:

→ **Granite schists**

⊖

Gneisses

30 a–h

from 22d

Gneiss is a textural concept that can be used in the field, which groups medium- to coarse-grained rocks that exhibit a coarse foliation formed by aligned micaceous minerals. The total amount of mica is not very large. The foliation can go from strongly developed to merely suggested. Gneisses also very often exhibit layered structures, sometimes even wavy, eye-shaped, and folded structures.
Different source rocks (protoliths) can be transformed to gneisses under high-grade metamorphic conditions. Gneisses can form via the metamorphosis of granitic rocks (granitic gneisses), or via the recrystallization of sandy-argillaceous sedimentary rocks (mica quartz gneisses to biotite plagioclase gneisses). Therefore two classes of gneisses can be distinguished based on the source rocks:

Orthogneisses → formed from igneous source rocks

Paragneisses → formed from sedimentary source rocks

Here we present a selection of the most common gneiss types. If you cannot find what you are looking for, you can simply list the attributes that seem important, as well as the recognizable minerals, and add them before the term "gneiss" to obtain a serviceable descriptive field designation.

Examples:
Light-colored, slabby, granite-like two-mica gneiss
Brown, flaser-like, quartz-rich biotite plagioclase gneiss

a Medium- to coarse-grained, light-colored gneiss that consists of feldspar, quartz, and micas (muscovite and/or biotite), just like granitic rocks. In many metamorphic bodies of granite the overprinting is selective and varies greatly. This is how batches with a predominantly plutonic structure, not yet altered to gneiss, transition to areas that are strongly altered to gneiss (and, depending on the tectono-metamorphic history, can also transition to granitic schists and all the way to granite-mylonite). A very common type of gneiss: → **Granite gneiss, granitic gneiss** ⊖

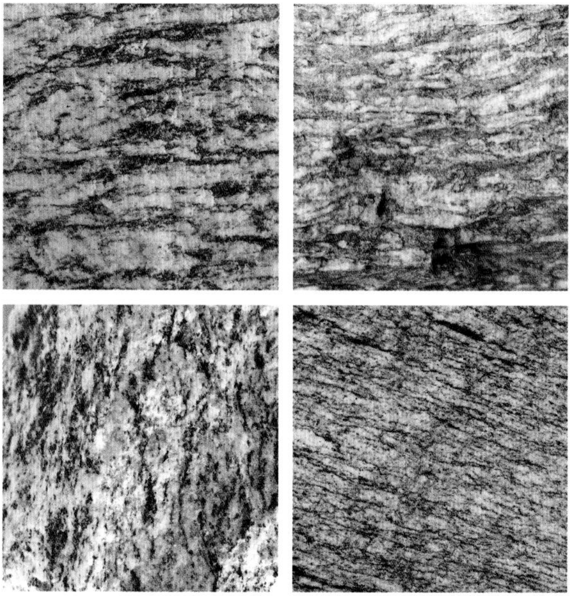

30

b Similar to (a), but with still-recognizable igneous porphyritic potassium feldspar phenocrysts. Metamorphic deformation may have created slightly "eyelike" shapes, and in extreme cases fine white layers. In linearly deformed augen gneisses the potassium feldspars are pulled apart to form acicular-like aggregates: → **Augen gneiss**

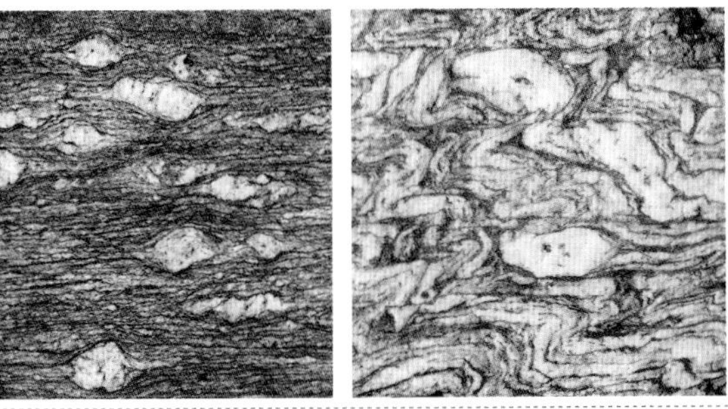

c Quartz-rich micaceous gneisses, which, aside from biotite and/or muscovite, consist of almost nothing but quartz, can develop from more or less argillaceous sandstones—a common sedimentary rock: → **Quartz-rich gneisses**

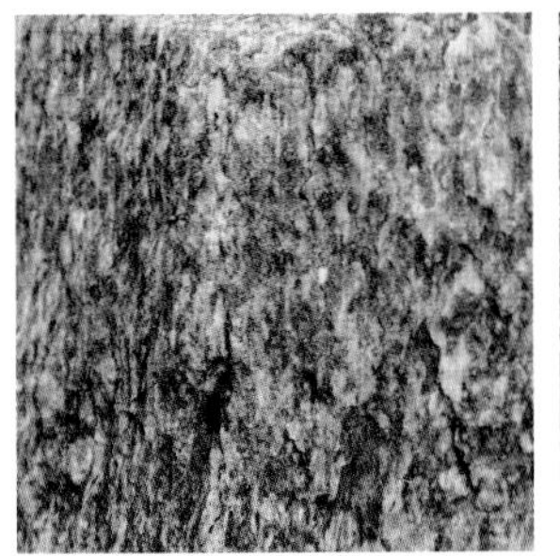

continued overleaf

Gneisses

30

Ch. / page

d Gray to brown gneisses consisting of biotite, feldspar (often plagioclase), and quartz are common in the crystalline bedrocks in many regions. They formed from sandstones that contained feldspar (graywackes, 7a, 21b):

→ **Biotite plagioclase gneisses** ⊖

e Gneisses with clear banding, mostly in lighter and darker layers. The banding can represent what was originally a bedding or it may arise from metamorphic differentiation:

→ **Banded gneiss** ⊖

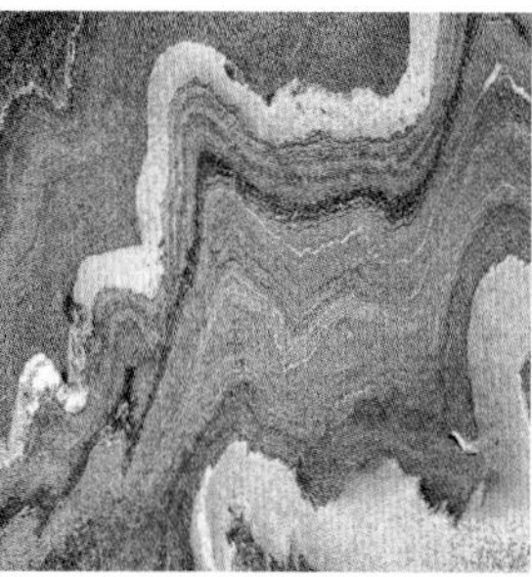

f Special case: in and of itself it is not classified among the gneisses.

→ **Amphibolite** (left)/ **banded amphibolite** (right) (see also 22e, 23c) ⊖ → 23c / 120

Dark or light blackish-green banded amphibolites can contain aligned biotites in their foliation/banding; the hornblendes can also be aligned along a number of planes:

g Gneiss-like white-green rock, basically devoid of mica, but often with the texture of augen gneiss. Green minerals = hornblende or actinolite; light areas = former plagioclase, transformed to a microcrystalline or fine-grained weave of albite, zoisite, etc.: → **Metagabbro/ flaser gabbro** ⊖

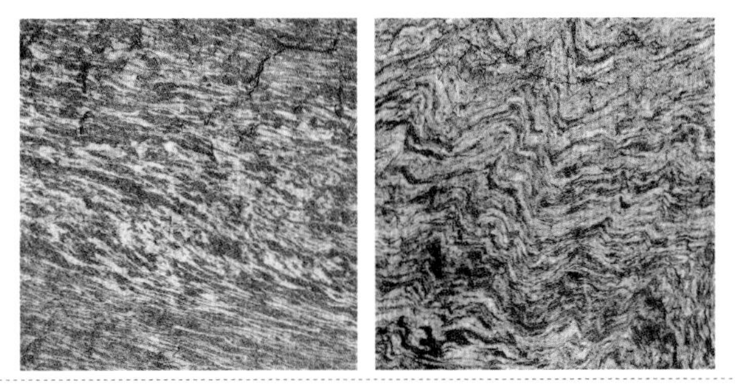

h Special case: in and of itself it is not classified among the gneisses → **Banded high-pressure rock** (see also 23e, 29f) ⊖

Both blueschists (29f) and eclogite (23e) can exhibit a more or less pronounced material banding or gneiss-like character. The typical high-pressure minerals (glaucophane or garnet omphacite), however, reveal these rocks as what they are:

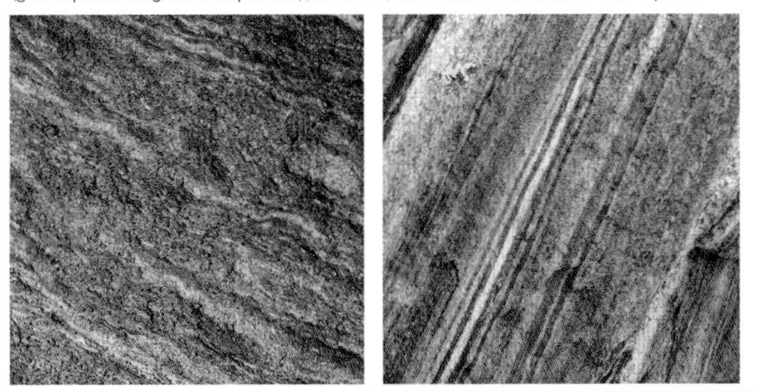

continued overleaf

Migmatites

31 a–c

from 22g

All kinds of gneiss can, at high temperatures of ca. 650–700° C and higher, become migmatites. Migmatites are hybrid rocks with a molten part and a gneissic remainder.
The melting of high-grade metamorphic gneisses begins first in veins, pockets, and patches. These normally exhibit a granitic composition (quartz and feldspar). These areas with an igneous texture are designated the "leucosome." The remaining gneiss is designated the "mesosome." Mostly at the boundary between the leucosome and mesosome a seam of dark minerals such as biotite and hornblende is formed, minerals that are "left over" during the fractional melting of the leucosome. These margins are designated the "melanosome."

a The mobilized, light-colored quartz feldspar areas (leucosomes) form layers that are cm to 10s of cm wide and run parallel to the foliation. They often exhibit a biotite-rich seam. Layered migmatites normally constitute the initial stage(s) of the melt: → **Layered migmatite** ⊖

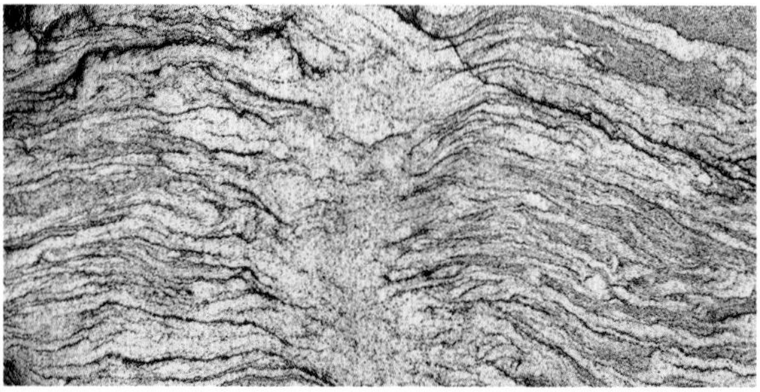

31

b The gneiss remains as suspended clods (angular or rounded) that are rotated against one another in the newly formed quartz feldspar matrix (leucosome). These kinds of migmatites bear witness to an advanced stage of melting:

→ **Schollen (raft) migmatite** ⊖

c A plutonic rock that in ghostlike fashion still allows recognition of its provenance from molten gneiss. This is mostly no longer recognizable in a hand specimen, but only at the level of a larger outcrop:

→ **Anatexite** ⊖

Ores

40 a–g

from 15 and 22h

Ores are a world unto themselves, which we can touch on only marginally here. Ores can make up 100% of a hand specimen (monomineralic) but can also be present in smaller quantities.
Massive ores exhibit a heavy to very heavy specific weight and consist mostly of metallic-lustrous, nontransparent minerals: the ore minerals. The most common are the following:

a Black, grainy, magnetic: → **Magnetite ore** ⊖

Magnetite, Fe_3O_4, is the most important iron ore.

b Black to reddish, often flaky, not magnetic: → **Hematite ore** ⊖

Hematite, Fe_2O_3, is also an important iron ore.

c Silver gray, very heavy, soft, cubic (blocky) fissility: → **Galena ore** ⊖

Galena, PbS, is the most important lead ore and often exhibits significant silver content.

40

d Goldish yellow, mostly running/iridescing into the colors of the spectrum: → **Chalcopyrite ore** ⊖

Yellow copper ore or chalcopyrite, $CuFeS_2$, is the most important copper ore.

e Brassy yellow, blocky grainy, very hard, often altered into rust-red Fe-hydroxide masses: → **Pyrite ore** ⊖

Pyrite, FeS_2, often also known as "fool's gold," was for a long time the most important raw material in the production of sulfuric acid.

f Brown to yellow, excellent fissility, strong luster: → Sphalerite / **Sphalerite ore** ⊖

Sphalerite or zincblende, ZnS, is the most important zinc ore; because of its often elevated cadmium content it is also the most important ore for this mineral.

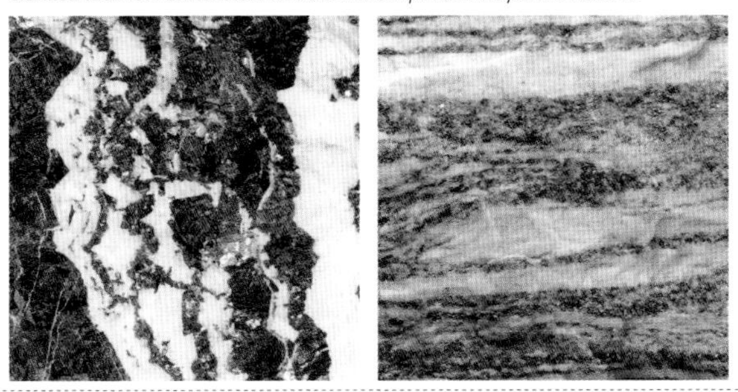

continued overleaf

Ores

40

g Black (possible minerals: manganite, hausmannite, manganese dioxide, and pyrolusite): → **Manganese ore**

These mostly hydrothermally formed ores are the most important raw materials for the extraction of manganese.

Glassy rocks

41 a–e

from 2e

Glassy rocks or rocks containing glass are formed mostly near volcanoes by the rapid cooling of lava, which prevents any crystals from forming so that the rock ends up consisting of cooled glassy magma.
Glassy rocks can, however, also develop in other locales with significant heat generation.

a Black volcanic glass, which often occurs in large chunks. Typically conchoidal glass-like fracture. The rock's composition is mostly "acidic" (rich in SiO_2, granitic). Occasionally it can contain areas with gas porosities or impurities: → **Obsidian**

41

b Brecciated volcanic rocks with pieces of volcanic glass in a very fine-grained matrix. Especially common in sequences of submarine pillow lavas as a result of explosive quenching of lava by seawater: → **Hyaloclastite**

c Black obsidian-like rock that runs through other rocks in irregular joints and veins, often with a brecciated structure. It is formed by very rapid tectonic movements at larger fault zones as a consequence of local frictional heat with temperatures above 1,000° C: → **Pseudotachylite** (see also 7g)

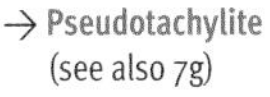

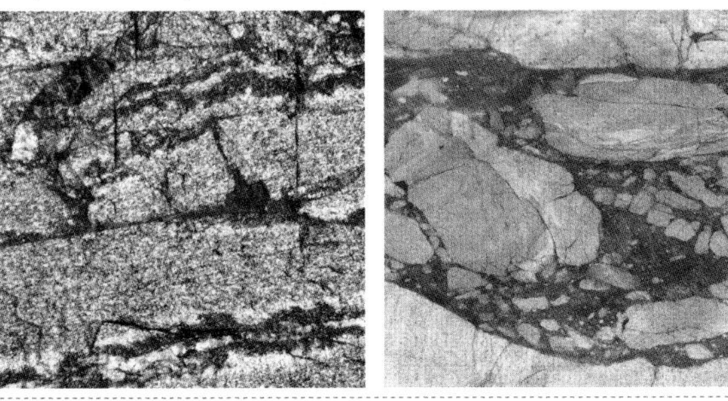

d Fine glassy coating, often bubble-like, primarily on top of high, exposed mountain peaks made of dark or iron-rich rocks. This glass is formed from local melting caused by lightning strikes. Fulgurites can also form in desert sands in the form of pipes made of welded quartz grains. → **Fulgurite**

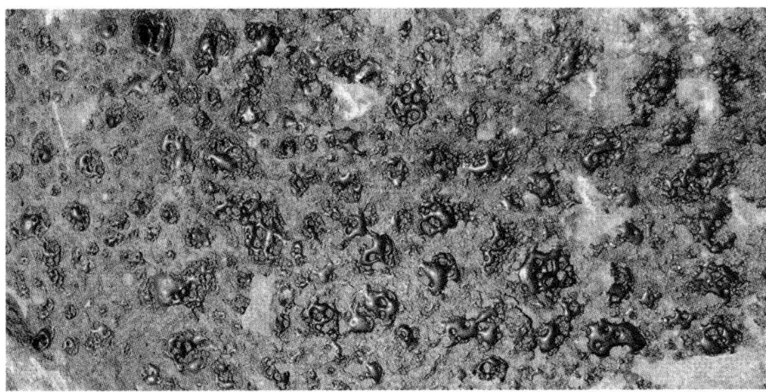

continued overleaf

Glassy rocks

41

e Glassy rocks up to several cm in size, black to green and translucent, formed by the impact of meteorites on the Earth's surface: → **Tektite** ⊖

Meteorite impact causes material from the Earth to become molten and be hurled up to several hundred kilometers away; there it solidifies to glass. The shapes can vary from aerodynamically round to very irregular. The green and translucent tektites of the Nördlinger Ries in Germany are called ***"moldavites."***

Strongly veined rocks

42 a–e

from 22f

Some rocks, ranging in size from hand specimens to outcrops, appear as vividly veined complexes, which often leave a chaotic impression. In most cases a combination of fragmentation because of brittleness and mineral precipitation from hot deep groundwater is responsible. These conditions predominate in many geological situations. Below is a selection of common possibilities.

a At oceanic spreading zones, with their many fault zones, intensive fragmentation of rocks can occur, reaching down to the peridotite layer. In these fault zones the seawater penetrates deeply into the crust: → **Ophicalcite/serpentinite breccia** ⊕

In this process peridotite is transformed into serpentinite. Vividly veined serpentinite rocks can be formed as a result. When these contain a lot of calcite in the form of joint fillings, they are referred to as ophicalcite. The rocks exhibit the green colors of serpentinite (11e, 12d), with greater or lesser quantities of white from the calcite.

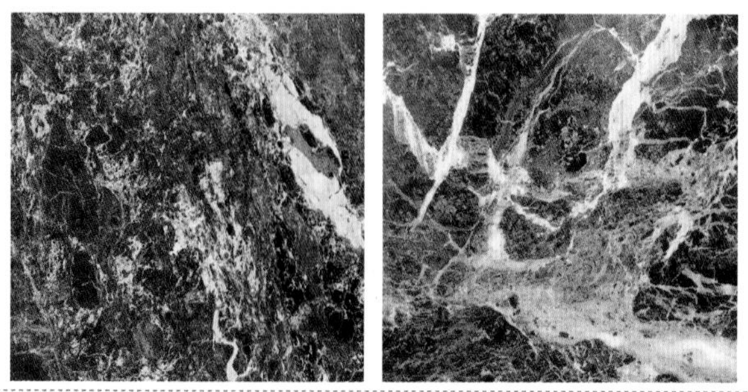

42

b Tectonic breccias of all kinds can contain a significant amount of cementing, as well as vein and joint fillings. Their mineral content often reflects the chemistry of the source rock. Ultimately these breccia rocks can be correctly identified only in their geological context:

→ **Tectonic breccia** (see also 7b, 44h) ⊕ ⊖

c In subterranean karst formations, cave-ins and collapses can occur, leading to the fracture of limestones and their later "healing" by precipitated calcite. Karst breccias always consist of limestones with calcite joints. Because of the oxidizing conditions, they are, however, often a strong rust red; they may also contain clay minerals and quartz sand grains.

→ **Karstic fissure filling** ⊕

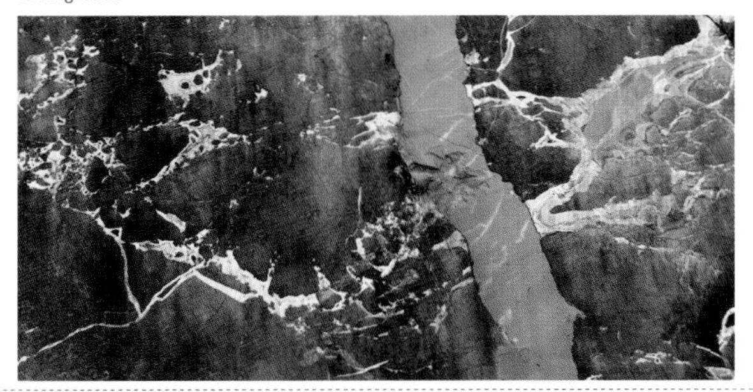

continued overleaf

Strongly veined rocks

42

Ch./page

d Hydrothermal veined rocks are ultimately no more than tectonic breccias from the greater depths of the crust, where mineral precipitation in the fissures occurs at higher temperatures. They are formed in volcanic substructures, around magma chambers, and near deep-reaching fault zones:

→ **Veined rocks** ⊕ ⊖ → 43/156

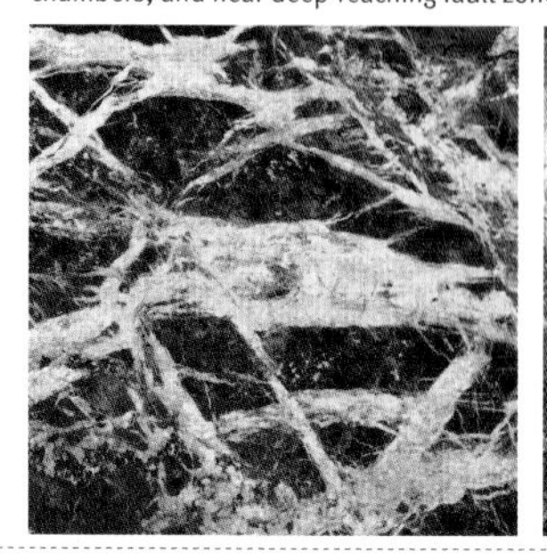

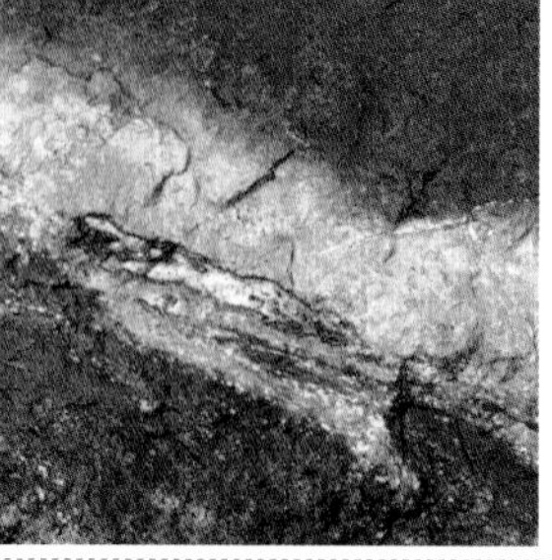

e Ore veins are a special case of either (b) or (d), where the crystallization in the veins or joints does not consist of common minerals, but rather of ore minerals. These can often be rather easily identified macroscopically:

→ **Ore veins** ⊕ ⊖

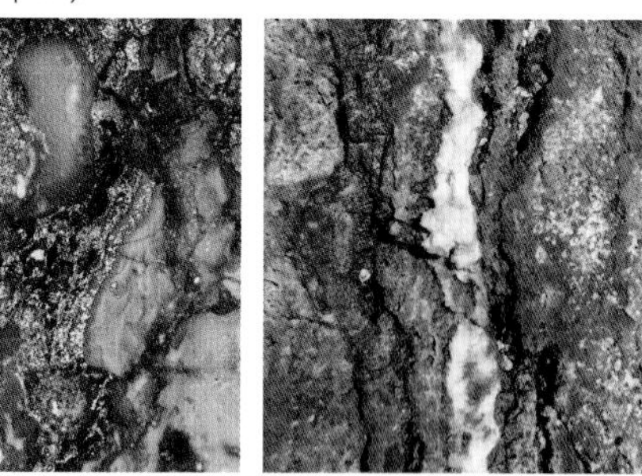

Vein fillings (of joints or fissures) → 43

Unconsolidated rock → 44

Vein fillings (of joints and fissures)

43 a–g

from 1b

Pieces of rock or pebbles often consist wholly of joint or vein fillings. Depending on the rock in which these fillings were formed, they can consist of a single mineral or all possible combinations. A number of these joint mineralizations appear frequently and in the most varied geological environments. In these cases the following rule of thumb applies:

The mineralogy (chemistry) of the rock determines the minerals in the veins!

So in limestone you will find calcite veins; in quartz sandstone, quartz veins; in granite, granite/quartz/(feldspar) veins; in greenschist, albite/(chlorite) veins, etc. The minerals in these joints often form larger grains that can be closely and chaotically interlocked; or, it is not rare to find acicular-like growth forms situated perpendicularly to the joint walls. Empty spaces with idiomorphically crystallized minerals are also possible.

a White, coarse, irregular fracture. Very hard (scratches steel and glass): → **Quartz veins** ⊖

b In addition to the white quartz there can also be white feldspar, which can usually be recognized thanks to its flat cleavage surfaces. Both are hard: → **Quartz feldspar veins** ⊖

43

c In addition to the white quartz there can also be calcite, which is usually recognizable thanks to its obliquely stepped cleavage surfaces (flat rhombohedra) and lack of hardness. The calcite is also often yellowish to rust brown: → **Quartz calcite veins** ⊕

d Mostly white feldspar, usually albite, together with calcite, which is usually recognizable thanks to its obliquely stepped cleavage surfaces (flat rhombohedra) and low hardness (H = 3). The calcite is also often yellowish to rust brown: → **Calcite feldspar veins** ⊕

e Pure calcite joints can be found in carbonate rocks. Their fissility and low hardness (H = 3) make them easily recognizable. Colors vary from white to rust red: → **Calcite veins** ⊕

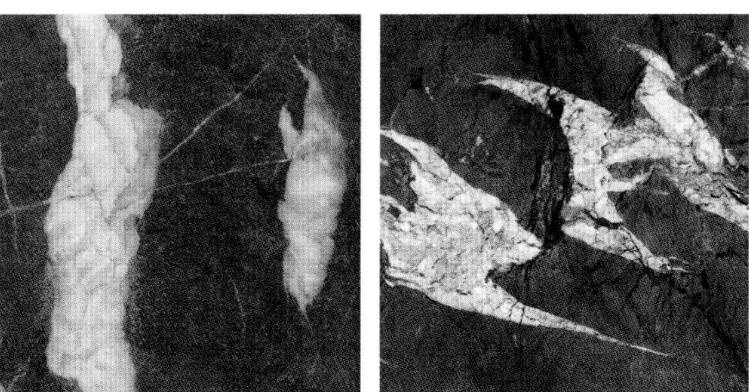

continued overleaf

Vein fillings (of joints and fissures)

43

f Pistachio-green coloration, either finely acicular or fine-grained aggregates, hard. Veins rich in epidote are common in igneous and metamorphic rocks that contain plagioclase feldspar and/or amphiboles and even pyroxenes. The epidote is often also combined with dull dark green chlorite and white albite feldspar:

→ **Epidote veins** ⊕ ⊖

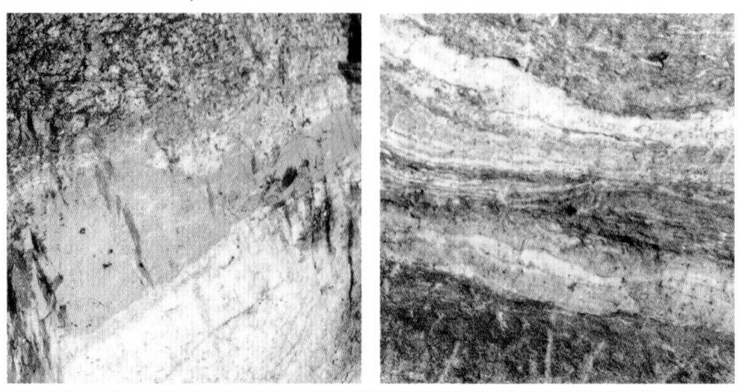

g In rocks similar to the epidote veins, as well as in serpentinites. Actinolite in veins occurs mostly in acicular and/or finely radiated aggregates. Light to rich green, acicular to fibrous.

→ **Actinolite veins** ⊖

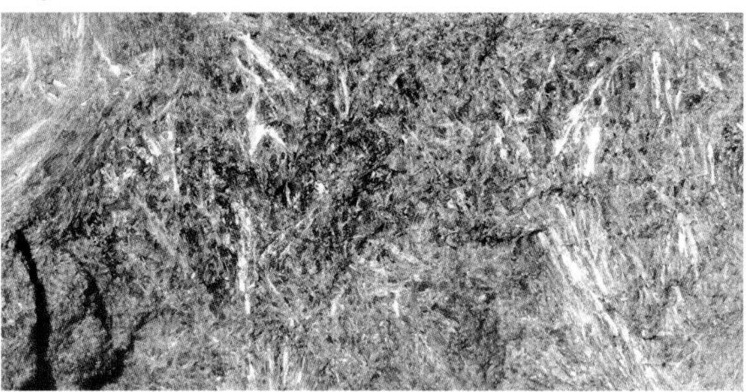

Unconsolidated rock

→ 44

Unconsolidated rocks

44 a–h

Ch. / page

from 1c

a Silt or Clay: Grain size smaller than 0.064 mm). → 45/163

b Sand: Grain size 0.064–2 mm. → 48/164

c Cobbles/Stones: Grain size 2–200 mm → 53/167

44

Ch./page

d Blocks: Grain size larger than 200 mm. → 54/168

e Brecciated-porous mass, ocher yellowish, can be really firm, but also brittle in the hand: → **Cornieule** see 4a ⊕ → 4a/68

f Earthy, rust-brown mass, sometimes homogeneous, sometimes heterogeneous, can also contain concretions and grains of quartz and be more or less argillaceous: → **Laterite** ⊖

continued overleaf

44

g Porous, earthy-soily mass, partially with still-recognizable plant remains: → **Peat** (actually more of a fossil soil). ⊖

h Breccia-like unconsolidated rock with directionless-chaotic structure from fault zones in the rock mass that arose because of significant tectonic stress: → **Fault gouge** ⊕ ⊖

Hardened/cemented fault gouges are referred to as → **tectonic breccias** (7b).

Very fine-grained unconsolidated rocks

45 a–b — Ch./page

from 44a

a When rubbed between the fingers while wet, feels slightly sandy. → 46/163

b When rubbed between the fingers while wet, feels slightly unctuous. → 47/163

Silt-like unconsolidated rocks

46

from 45a

a Narrowing down the identification in the geological context is necessary. → **Silt/loess** (left) **volcanic tuff** (right) ⊕

Clays and clay-like unconsolidated rocks

47 a–c

from 45b

a Unstructured, very fine-grained mass; colors variable, mostly gray, but also very light beige to almost white, also rust red. Swells in volume when wet, becomes kneadable: → **Clay** ⊖

continued overleaf

Clays and clay-like unconsolidated rocks

47

b Very fine, light-colored mass: → **Marine or limnic lime sludge** ⊕

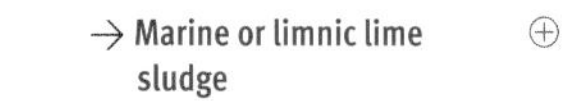

c Whitish, powdery: → **Diatomite, diatomaceous earth,** → = unhardened, Quaternary marine or lacustrine diatomaceous sludge. ⊖

Sands

48 a–d

from 44b

		Ch./page
a	Sand grains very hard (scratch steel) and ± colorless, transparent.	→ 49/165
b	Sand grains hard to very hard (mostly scratch steel), various color tones.	→ 50/165
c	Sand grains hard to very hard (mostly scratch steel), with bold colors (black, red, yellow).	→ 51/166
d	Sand grains soft (do not scratch steel or copper blade), dull whitish or light color tones.	→ 52/166

Quartz sand

49

from 48a

→ **Quartz sand** Pure quartz sands can occur in rivers, but mostly near seacoasts/deltas (left photo) or in deserts (right photo).

Silicate sand

50

from 48b

→ **Sand from quartz and feldspar** (and occasionally from other silicate minerals).

Special sands

51

from 48c

→ **Special sands** ⊖
like garnet sand, for example (right photo), olivine sand, magnetite sand, or mixtures of different silicate and ore minerals (left photo).

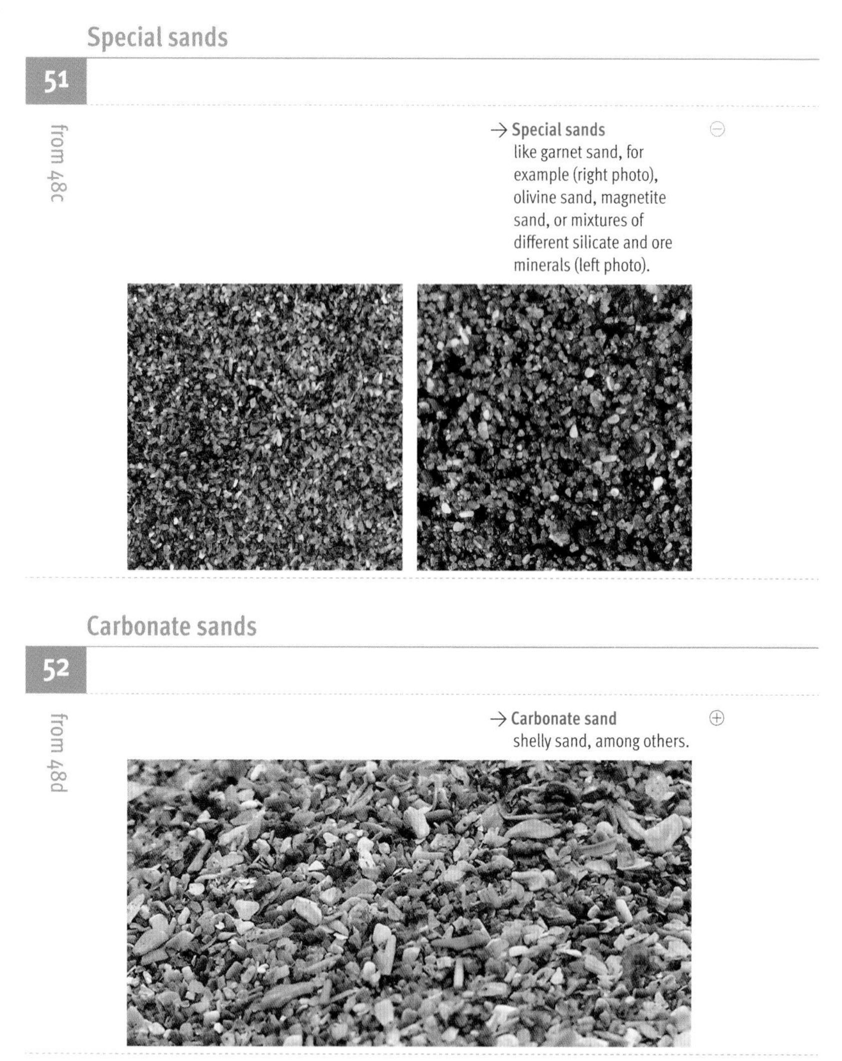

Carbonate sands

52

from 48d

→ **Carbonate sand** ⊕
shelly sand, among others.

Unconsolidated rocks in the stone (cobble) grain-size range

53 a–c — Ch./page

from 44c

a Rounded stones, pebbles: → **Gravel** (mostly river gravel) ⊕ ⊖ → 55/169

b Angular stones: → **Hillside scree** (rarely fault breccia) ⊕ ⊖

c Angular but also rounded stones lie in a fine, often also porous matrix. The components are often bubbly: ⊖

A Components are angular: → **Volcanic tephra**

B Components are rounded: → **Volcanic lapilli**

Unconsolidated rocks consisting of blocks

54 a–c

from 44d

a Rounded blocks: → **Beach blocks** ⊕ ⊖

b A mixture of blocks and stones (cobbles), only partially rounded, often in a silty (or also slightly argillaceous) matrix: → **Moraine (Till, diamictite)** ⊕ ⊖

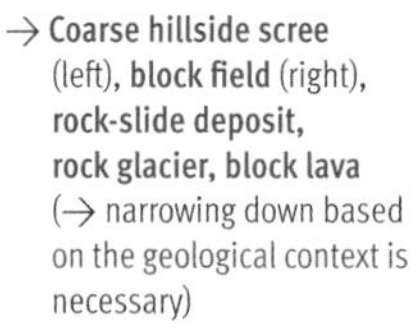

c Angular blocks: → **Coarse hillside scree** (left), **block field** (right), **rock-slide deposit, rock glacier, block lava** (→ narrowing down based on the geological context is necessary) ⊕ ⊖

Gravels

55 a–b

from 53a

a By a stream or river: → **River gravel** ⊕ ⊖

River gravels from more recent geological ages (Quaternary) are important raw materials for the preparation of concrete, gravel beds, and so on. Deposits in central Europe are usually from the glacial periods.

b By an ocean beach: → **Beach gravel** ⊕ ⊖

Rocks not dealt with in the identification key

The following rocks are not included in the identification key; they are presented here very briefly in alphabetical order. There are additional very exotic or rare kinds of rocks that are also not dealt with in this book—in these cases you should refer to a specialist or to the relevant literature.

Banded Iron Formation

Clearly banded rock formations. Rocks built in layers of chert and iron ore (magnetite, hematite) mm to cm thick. The Banded Iron Formation was formed exclusively in the Archean and Proterozoic eons, presumably in connection with the production of oxygen by microorganisms. It is present in the old continental shields of South Africa, Australia, and Russia as well as South and North America. It contains the most significant iron deposits in the world.

Beachrock

Beach deposits hardened on location by means of carbonate binders on tropical and subtropical beaches; forms hard, flat-lying rock "lids" in beach areas.

Charnockite

This is a collective term for quartz feldspar rocks with a plutonic structure that contain orthopyroxenes and that occur mostly in Proterozoic metamorphic rock sequences. They are today considered to be special metamorphic rocks of the granulite facies. Considerable amounts in India, Brazil, and Nigeria are used as construction and decoration material.

Fenite

Coarse-grained rocks with alkali feldspars, natrium amphiboles and pyroxenes, titanite, and apatite. They are formed around large alkali intrusions into a variety of surrounding rocks (= fenitization) because of the influence of temperature and alkaline liquids. The fenite aureoles can extend several km. Related to the skarns.

Greisen

The concept derives from the jargon of Saxon miners. Gray, quartz-rich, granite-like rocks, formed around granite plutons because of the influence of igneous fluids (rich in fluorine, lithium, tungsten, and zinc, among others).

Carbonatite	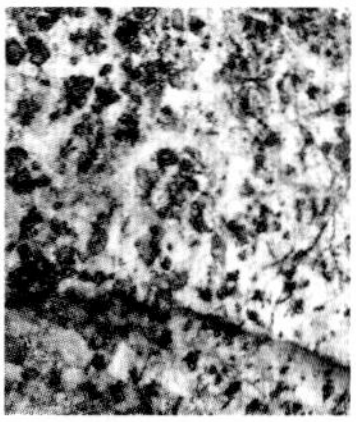 Very rare igneous rocks that consist of more than 50% carbonate minerals (most frequently calcite). They are present in combination with SiO_2-poor magmas. Extremely rare are carbonatitic volcanic rocks; the most renowned is the active Ol Doinyo Lengai volcano in Tanzania.
Kimberlite	 Ultramafic plutonic-volcanic rocks that are expelled directly from the Earth's mantle through vents. The vent fillings are shattered to breccias because of the explosive emissions, and the rocks are serpentinized. Kimberlites often contain diamonds. They occur in Africa, Australia, Siberia, Brazil, and Canada.
Komatiite	 Ultramafic volcanic rocks named after the Komati River in South Africa. They occur almost exclusively in the greenstone belts of the Archean continental shields. More recent Upper Cretaceous komatiites have been found in the Pacific plate.
Phosphorite	 Phosphorites are mostly nodule-like formations composed of fluorapatite and calcite. They occur in marine mudrocks and limestones and usually mark condensation horizons, witnesses of long sedimentation interruptions.
Suevit	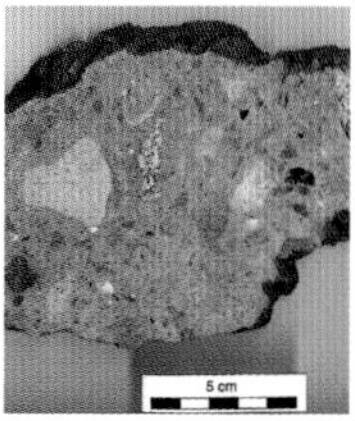 Rock formed by the impact of a large meteorite. These are impact breccias that can contain rock debris, glass, and high-pressure modifications of quartz (stishovite, coesite). Suevite was originally described based on the corresponding impact rocks of the Nördlinger Ries in Germany.

Meteorites: Messengers from the solar system

Meteorites—a large, exciting, and fascinating universe of rocks! Meteorites are chunks of rock from our solar system that collided with the Earth and survived the descent through the atmosphere, at least partially. There are both metallic and stony meteorites.
We will not examine meteorites further here, as it would be outside the scope of this volume. There is an abundance of good literature on the topic.

OPPOSITE PAGE:
Intrusive contact of fine-grained granite from the Bergell pluton (Swiss-Italian Alps), with high-grade metamorphic gneisses at the eastern edge of the pluton, and small intrusions into the gneisses.

PART 3

an introduction to rock classification

Igneous Rocks 1: Plutonic Rocks (Plutonites)

Two diagrams can be used to classify igneous rocks:

1. A diagram showing the amounts of the most important minerals (fig. below) is easy to utilize but is almost too simple. So, for instance, most granites do not need to contain hornblende—not by a long shot, as can be seen from the diagram. This is true for plutonites and volcanics.
2. The "Streckeisen diagram," named after Albert Streckeisen (1901–1998), is also known as the QAPF diagram (next page). There is one diagram for plutonites and another for volcanics. It is based only on the light-colored (leucocratic) minerals quartz, alkali feldspar, plagioclase, and foids (feldspathoids). The dark-colored (mafic) minerals biotite (and muscovite), hornblende, pyroxenes, and olivine are used only as additional elements in naming. The diagram is valid for all plutonites that contain at least 10% light minerals by volume. So even a very dark hornblende biotite diorite that contains only about 20% plagioclase will still be classified solely on this basis. But the diagram works well. A more precise classification is possible only if you can keep the two feldspars, potassium feldspar and plagioclase, separate. For ultramafic rocks with more than 90% dark mineral content by volume, some additional diagrams come into play.

Aplite (25a), pegmatite (28a), and lamprophyre (13b), which are defined as dike rocks based on their structure and geological context in addition to their mineral content, also belong to the plutonites.

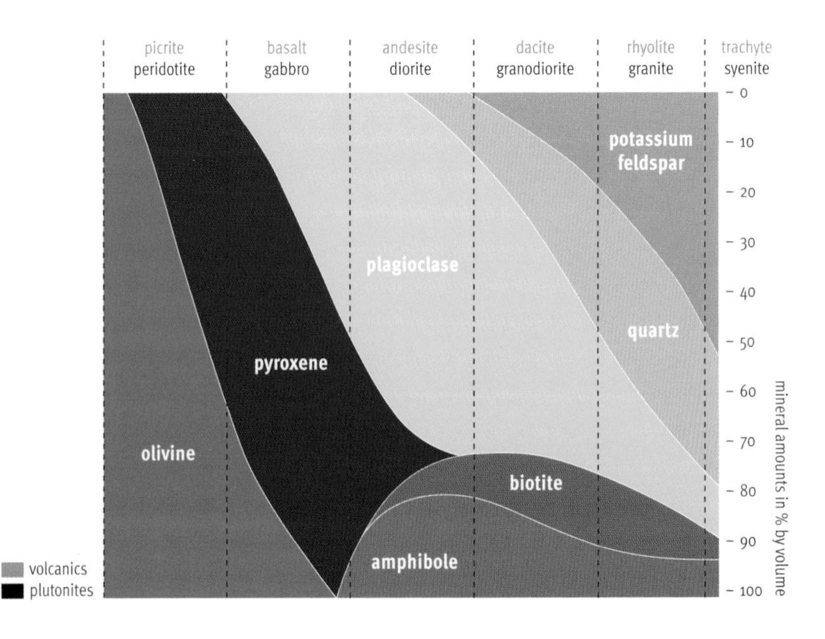

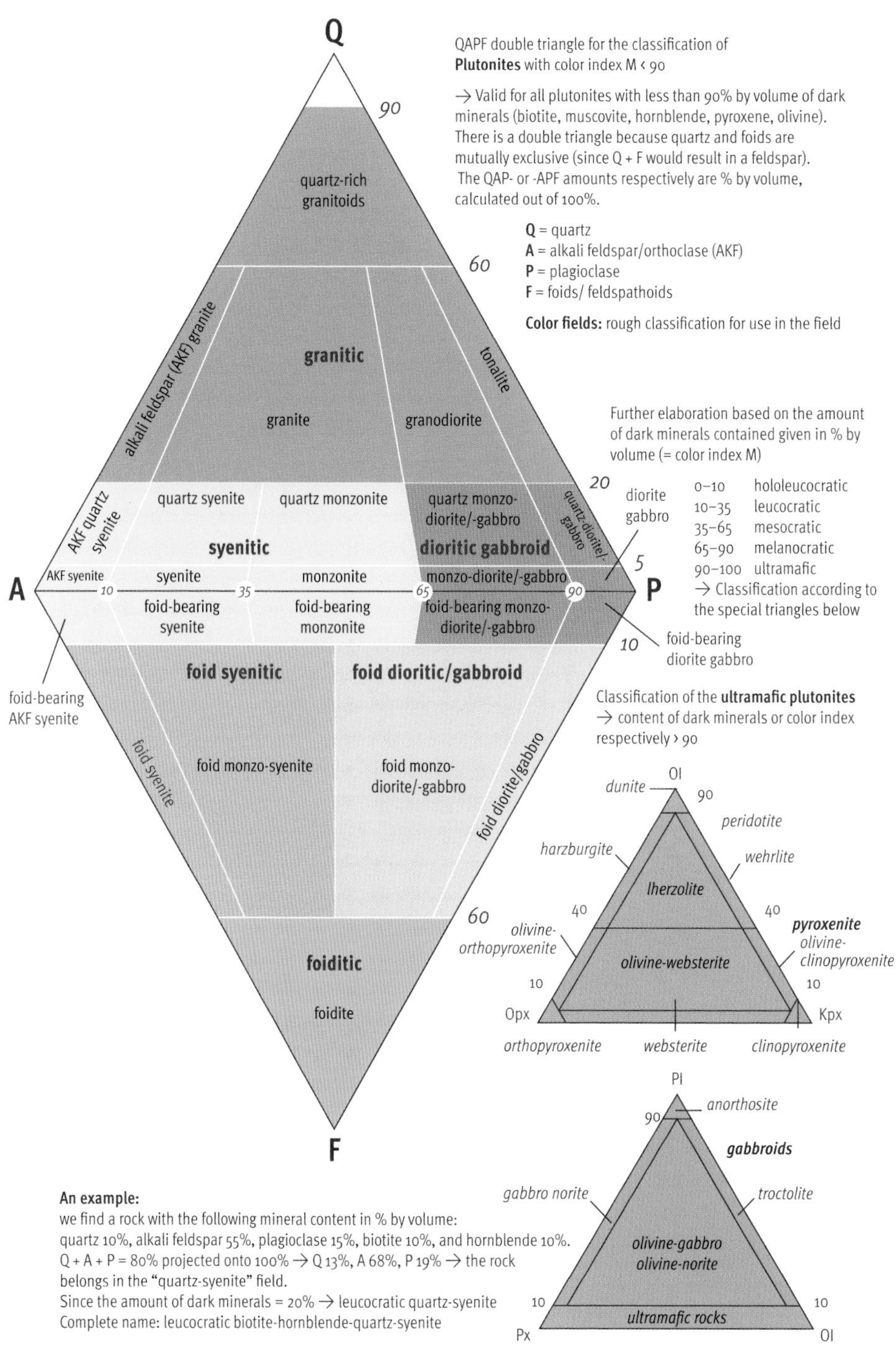

QAPF double triangle for the classification of **Plutonites** with color index M < 90

→ Valid for all plutonites with less than 90% by volume of dark minerals (biotite, muscovite, hornblende, pyroxene, olivine). There is a double triangle because quartz and foids are mutually exclusive (since Q + F would result in a feldspar). The QAP- or -APF amounts respectively are % by volume, calculated out of 100%.

Q = quartz
A = alkali feldspar/orthoclase (AKF)
P = plagioclase
F = foids/ feldspathoids

Color fields: rough classification for use in the field

Further elaboration based on the amount of dark minerals contained given in % by volume (= color index M)

0–10	hololeucocratic
10–35	leucocratic
35–65	mesocratic
65–90	melanocratic
90–100	ultramafic

→ Classification according to the special triangles below

Classification of the **ultramafic plutonites** → content of dark minerals or color index respectively > 90

An example:
we find a rock with the following mineral content in % by volume:
quartz 10%, alkali feldspar 55%, plagioclase 15%, biotite 10%, and hornblende 10%.
Q + A + P = 80% projected onto 100% → Q 13%, A 68%, P 19% → the rock belongs in the “quartz-syenite” field.
Since the amount of dark minerals = 20% → leucocratic quartz-syenite
Complete name: leucocratic biotite-hornblende-quartz-syenite

Igneous Rocks 2: Volcanic Rocks

Volcanics can also basically be classified with the help of the Streckeisen diagram (p. 177). However, this is often not possible because the minerals cannot be identified with a magnifying glass. In addition, we are dealing with products of volcanic activity, which lead to quite distinctive kinds of rocks like tuffs (46), lapilli (4c, 53c), or ignimbrites (13c). These can be present as both loose and solid rock. Volcanics can be additionally described based on the forms they acquired in the outpouring they were part of, as in basalts, the aa or pahoehoe lavas, or, where they have flowed out below the ocean, pillow lavas. A brief overview in alphabetical order can be found below.

Ashes	Very fine-grained tephra with grain sizes below 2 mm. Subsequent to deposition it is frequently rapidly transformed into clay minerals. These "bentonite horizons" can help document large volcanic eruptions even thousands of kilometers from the volcanic vent.
Pumice	Foamed-up SiO_2-rich lava, light gray to white, solidified in glassy forms.
Bombs (lava balls)/blocks	Tephra material(s) over 64 mm; bombs exhibit plastic trajectory deformations.
Ignimbrite (or welded tuff)	More or less heavy to hard-dense to porous rock, solidified deposit of hot pyroclastic flows out of which molten particles were "welded" together during deposition.
Lahar	Water-rich volcanic detritus or mud lava flows that result in chaotic brecciated deposits.
Lapilli	Tephra materials of ca. 2–64 mm.
Pyroclastite	Umbrella term for clastic (composed of fragments) volcanic rocks.
Tephra	Umbrella term for all loose volcanic materials formed from pyroclastic fall deposits and unconsolidated deposits.
Tuff	Solidified tephra (ash tuff, lapilli tuff).

Both plutonites and volcanics are formed by the cooling of magma. The question therefore arises about how plutonites can be distinguished from volcanics: When exactly does a plutonite transition to a volcanic? In the immediate substructure of volcanoes—in their vents, fissure fillings, and so forth—rocks can crystallize that exhibit characteristics of both groups. As is so often the case in the natural sciences, an attempt to make the problem less acute created an intermediate grouping: the subvolcanic rocks. They develop in the superstructure of plutons or the substructure of volcanoes, at depths ranging from a few hundred meters to a few kilometers. While this measure has proven practically useful, we now naturally have to define two demarcations instead of one.

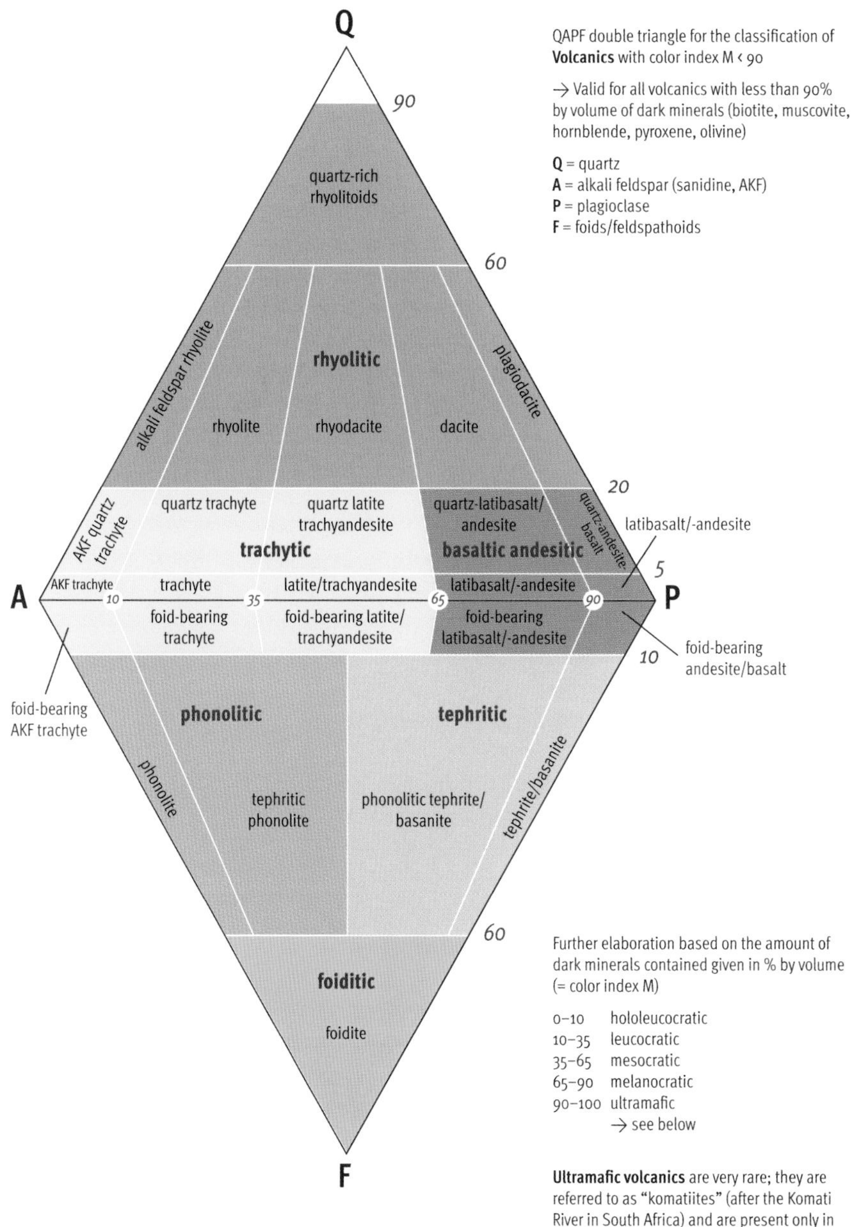

QAPF double triangle for the classification of **Volcanics** with color index M < 90

→ Valid for all volcanics with less than 90% by volume of dark minerals (biotite, muscovite, hornblende, pyroxene, olivine)

Q = quartz
A = alkali feldspar (sanidine, AKF)
P = plagioclase
F = foids/feldspathoids

Further elaboration based on the amount of dark minerals contained given in % by volume (= color index M)

0–10	hololeucocratic
10–35	leucocratic
35–65	mesocratic
65–90	melanocratic
90–100	ultramafic → see below

Ultramafic volcanics are very rare; they are referred to as “komatiites” (after the Komati River in South Africa) and are present only in Archean continental shields.

Sedimentary Rocks

Sedimentary rocks are divided into three types: clastic, biogenic (organic), and chemical.

Clastic Sediments

These are classified according to their grain size (fig. next page): stones and gravels (⇢ conglomerates); sand (⇢ sandstones); silt (fine sandstones, loess); and clay (mudrock, pelites), as well as according to their depositional environment (terrestrial, fluvial, glacial, intertidal, marine). So, for instance, pure quartz sandstone may have been deposited terrestrially in a desert, fluvially on a river's sandbank, or in a marine environment in a delta or a turbid current. To proceed with identification, the geological context, specific structures, or fossils are usually necessary. This is why this identification key does not proceed further into the genesis of these deposits. The same qualifications are basically valid for breccias, which can have extremely diverse origins.

Biogenic (Organic) Sediments

Carbonate rocks that formed in a marine environment occupy the foreground in this category—flat marine areas could be seen as a "carbonate factory," and fine-grained micritic limestones could be viewed as gigantic "plankton cemeteries." There are two further classifications of carbonate rocks: Folk's system fundamentally distinguishes between the kinds of carbonate particles (intraclasts, ooids, bioclasts, and pellets) and the "matrix" between them (lime sludge or calcareous cement), while Dunham's classification is instead based on the structures resulting from different energy flows.

In the identification key we present the most important macroscopically distinguishable types of limestone: micritic (12a), oolitic (17c), spathic (17a), and bioclastic (17d).

A special case of biogenic (organic) sedimentary rocks is represented by the radiolarites, which arose in the deep ocean from radiolaria and their quartz skeletons.

Grain sizes	Blocks	Stones (cobbles)	Gravel			Sand			Silt	Clay
			coarse	medium	fine	coarse	medium	fine		
Grain sizes in mm	> 200	63 – 200	63 – 20	20 – 6.3	6.3 – 2	2 – 0.63	0.63 – 0.2	0.2 – 0.06	0.06 – 0.002	< 0.002
Size comparisons	Watermelon	Melon	Chicken egg	Hazelnut	Pea	Match head	Wheat semolina	Powdered sugar	Can be kneaded	Can be rolled out
Unconsolidated rocks	Boulder fields	Scree fields, hillside scree	Fluvial or beach gravels			River, coastal/delta, and desert sands			Loess	Clay
Solid rocks	Breccias, fanglomerates, and beach conglomerates		Conglomerates/breccias			Sandstones (including graywacke, arkoses)			Marl	Mudrock

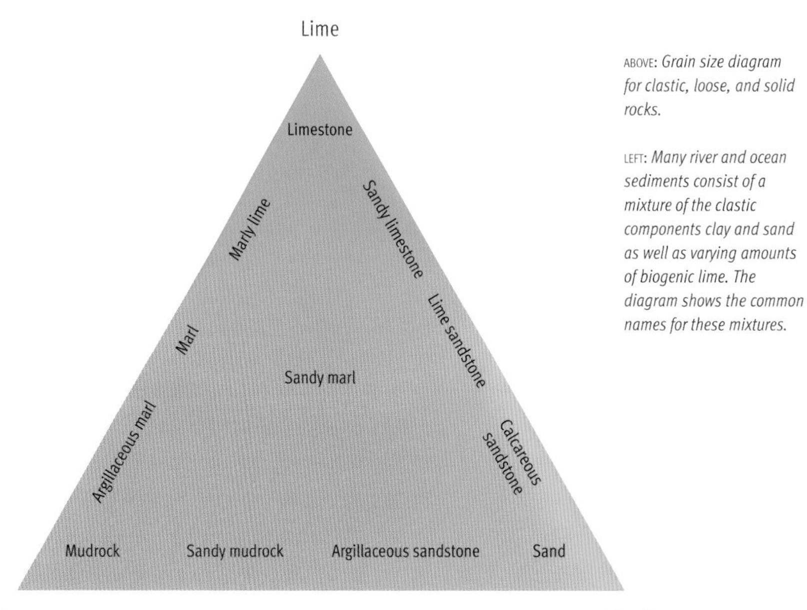

ABOVE: *Grain size diagram for clastic, loose, and solid rocks.*

LEFT: *Many river and ocean sediments consist of a mixture of the clastic components clay and sand as well as varying amounts of biogenic lime. The diagram shows the common names for these mixtures.*

Metamorphic Rocks

The nomenclature for metamorphic rocks is—sigh!—a pretty mixed bag. Classification into different pressure and temperature categories, the so-called metamorphic facies, was defined on the basis of metamorphic basalt rocks. Why? Very simple: metabasalts react sensitively to different pressures and temperatures and also form rock types that are easy to distinguish macroscopically—such as greenschist, blueschist, amphibolite, and eclogite.

So all metamorphic rocks can be described with the corresponding facies concept and the prefix "meta"—as long as you know the source rock. Here are three examples of this "genetic" naming: amphibolite facies metagranite / greenschist facies metapelite / eclogite facies metabasalt.

Second, the three structural concepts "schist," "gneiss," and "fels" are combined with the most important minerals present in the rock. So the above examples, following this descriptive nomenclature, could become two-mica granite gneiss / chlorite-biotite-sericite schist / kyanite-bearing omphacite garnet fels.

Third, in the course of history different metamorphic rocks were assigned special proper names, and you simply have to become familiar with them. For the first and third examples above we then have the designations orthogneiss and eclogite.

The International Union of Geological Sciences (IUGS, www.iugs.org) published a recommendation in 2007 with uniform usage of the entire nomenclature, which we are using here. On the next page is a facies diagram, and on the following pages a large table allows you to understand which metamorphic rocks originate from which source rocks given the different facies.

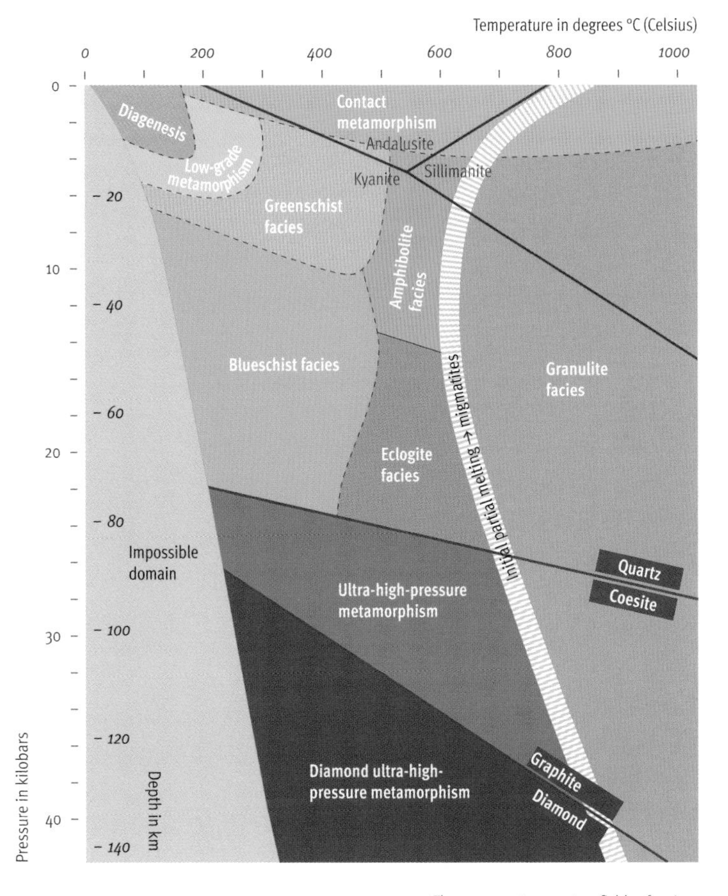

The pressure-temperature fields of various metamorphisms (= metamorphic facies)

Source rocks and their metamorphic products with different degrees of metamorphism (metamorphic facies, see fig. p. 181)

In each box in the table the name of the rock is given above, while below you will find the most important metamorphically newly formed (or recrystallized) minerals contained in them.

Premetamorphic rocks (protolith) and mineralogy	Observations	Regional metamorphism (Barrow type)		
		Weakly metamorphic	**Greenschist facies**	**Amphibolite facies**
Limestone (dolomite) Calcite (dolomite)	Since it consists of only one mineral, calcite (or dolomite) → only recrystallizations.	**Limestone (dolomite)** Calcite (dolomite)	**Calcite- (dolomite-) marble** Calcite (dolomite)	**Calcite- (dolomite-) marble** Calcite (dolomite)
Quartz(ose) sandstone Quartz, possibly some argillaceous minerals	Since it consists of basically one mineral, quartz ⇢ only recrystallizations (possibly some white mica formation → micaceous quartzite).	**Sandstone** Quartz (possibly some argillaceous minerals)	**Quartzite** Quartz (possibly some white mica)	**Quartzite** Quartz (possibly some white mica)
Mudrock (pelite) Various argillaceous minerals, possibly some quartz, a little carbonate	Chemically complex, react(s) with numerous new minerals with different degrees of metamorphism.	**Argillaceous shale** Illite (from argillaceous minerals), chlorite	**Slate / roof slate** Sericite (white mica), pyrophillite, chlorite, chloritoid, biotite	**Mica schist** Mica, feldspars, quartz, garnet, staurolite, kyanite
Marl Calcite or dolomite, argillaceous minerals, quartz	Chemically complex, react(s) with numerous new minerals with different degrees of metamorphism.	**Marly shale** Illite, possibly chlorite, ankerite	**Calcareous slate** Sericite (white mica), chlorite, albite, quartz, calcite	**Calc-silicate rock** Calcite, quartz, feldspar, biotite, garnet, hornblende, vesuvianite, diopside
Graywacke or impure sandstone Quartz, feldspars, mica, etc	Common rocks in older Paleozoic and earlier rock sequences, which were transformed into large amounts of biotite-plagioclase gneisses	**Metagraywacke** Zeolite, prehnite, pumpellyite, illite	**Sericite-chlorite-quartz schist** Sericite (white mica), chlorite, albite, epidote, actinolite	**Biotite-plagioclase gneiss** Plagioclase, quartz, biotite, hornblende, possibly garnet

		High-pressure metamorphism (subduction)			Contact metamorphism	
	Granulite facies	**Blueschist facies**	**Eclogite facies**	**Ultra-high-pressure metamorphism**	**Contact metamorphism weak**	**Contact metamorphism strong**
	Calcite-(dolomite-) marble	**Calcite-(dolomite-) marble**	**Calcite-(dolomite-) marble**	**Calcite-(dolomite-) marble**	**Limestone (dolomite)**	**Calcite-(dolomite-) marble**
	Calcite (dolomite)	Calcite (dolomite)	Calcite (dolomite)	Calcite (dolomite)	Calcite (dolomite)	Calcite (dolomite)
	Quartzite	**Quartzite**	**Quartzite**	**Quartzite**	**Sandstone**	**Quartzite**
	Quartz (possibly some white mica)	Quartz (possibly some white mica)	Quartz (possibly some white mica)	Quartz (possibly some white mica)	Quartz (possibly some argillaceous minerals)	Quartz (possibly some feldspar, sillimanite, garnet)
	Metapelitic granulite	**Slate / roof slate**	**Eclogitic mica schist**	**Whiteschist**	**Andalusite-slate, maculose (knotenschiefer)**	**Hornfels**
	Garnet, orthopyroxenes, cordierite, sillimanite, sapphirine, biotite, spinel	Sericite (white mica), chlorite, albite, carpolite, quartz	Quartz, white mica, garnet, kyanite, talc, chloritoid	Coesite, white mica, talc, kyanite, Mg-chloritoid, garnet (pyrope)	Andalusite, muscovite, chlorite	Feldspars, quartz, cordierite, sillimanite, garnet
	Calc-silicate rock	**Calcareous slate**	**Calc-silicate rock**	**Calc-silicate rock**	**Metamarl**	**Calc-silicate-hornfels**
	Calcite, quartz, feldspar, garnet, vesuvianite, diopside, wollastonite, corundum	Sericite (white mica), chlorite, albite, quartz, calcite, carpolite, glaucophane	Calcite, coesite, omphacite, garnet, white mica	Calcite, coesite, omphacite, garnet, white mica	White mica, chlorite, clinozoisite	Feldspars, clinopyroxene, wollastonite, garnet
	Granulite	**Sericite-chlorite-quartz schist**	**Omphacite gneiss**	**Omphacite gneiss**	**Metagraywacke**	**Graywacke-hornfels**
	Quartz, feldspars, garnet, pyroxenes, biotite, sillimanite	Sericite (white mica), chlorite, albite, quartz, zoisite, glaucophane, lawsonite	Omphacite, quartz, white mica, zoisite, garnet	Omphacite, coesite, white mica, zoisite, garnet	Prehnite, pumpellyite, illite	Feldspars, pyroxene, garnet, cordierite, sillimanite

continued overleaf

continued

Source rocks and their metamorphic products with different degrees of metamorphism (metamorphic facies, see fig. p. 181)

In each box in the table the name of the rock is given above, while below you will find the most important metamorphically newly formed (or recrystallized) minerals contained in them.

Premetamorphic rocks (protolith) and mineralogy	Observations	Regional metamorphism (Barrow type)		
		Weakly metamorphic	Greenschist facies	Amphibolite facies
Basalt (gabbro/ diorite Plagioclase, pyroxene, possibly olivine, with diorite, also hornblende, biotite	Chemically complex, react(s) with numerous new minerals with different degrees of metamorphism. Metamorphic basalts are eponymous for the metamorphic facies! In the case of gabbros, coarse schieren-like structures are easily formed (→ flaser gabbro)	**Metabasalt** Zeolite, prehnite, pumpellyite, calcite	**Greenschist/ greenstone** Albite, chlorite, actinolite, epidote	**Amphibolite** Plagioclase, hornblende, possibly garnet, epidote, biotite
Granite Feldspar, quartz, mica	In the case of deeper metamorphic degrees, incomplete transformations frequently occur, depending on deformation and fluid access. With higher degrees of metamorphism, basically the same mineral composition as the igneous protolith, recognizable as a metamorphic rock mostly thanks to structure → gneiss, augen gneiss	**Granite** Saussuritization/ "clayification" of feldspars	**Greened granite** Saussuritization/ micatization of feldspars, chlorite, epidote, stilpnomelane	**Orthogneiss (granite gneiss)** Feldspar, quartz, mica (recrystallized)
Peridotite Olivine, possibly pyroxenes, phlogopite	Chemically simple, it is especially influenced by water influx in the oceanic crust, as a result of which olivine is transformed into the "olivine + water" mineral serpentine. At high temperatures it once again recrystallizes to olivine.	**Serpentinite** Chrysotile, lizardite, brucite	**Serpentinite** Antigorite, brucite, talc, chlorite	**Serpentinite** Antigorite, possibly anthophyllite, titanian clinohumite, magnesite, olivine

	High-pressure metamorphism (subduction)			Contact metamorphism	
Granulite facies	**Blueschist facies**	**Eclogite facies**	**Ultra-high-pressure metamorphism**	**Contact metamorphism weak**	**Contact metamorphism strong**
Granulite, basic	Blueschist	Eclogite	Eclogite	Metabasalt	Amphibolite
Plagioclase, ortho- and clino-pyroxene, garnet	Glaucophane, lawsonite, epidote, white mica, chlorite	Omphacite, garnet, possibly mica, kyanite, talc	Omphacite, garnet, possibly mica, kyanite	Zeolite, prehnite, pumpellyite	Plagioclase, hornblende, possibly garnet, epidote
Orthogneiss (granite gneiss)	Metagranite to orthogneiss	Orthogneiss (granite gneiss)	Orthogneiss (granite gneiss)	Granite	Granite
Feldspar, quartz, garnet, hornblende, pyroxene, biotite	Feldspar, quartz, mica (recrystallized)	Jadeite, zoisite, quartz, omphacite, mica	Jadeite, zoisite, coesite, omphacite, mica	Saussuritization/ "clayification" of feldspars	Feldspar, quartz, mica remain preserved
Olivine fels	Serpentinite	Serpentinite	Garnet-peridotite	Peridotite	Peridotite
Olivine, ortho-/ clino-pyroxene, garnet, spinel	Chrysotile, lizardite	Antigorite	Olivine, pyrope-garnet, diopside	Serpentine minerals (only with the addition of water)	Olivine, pyroxene, anthophyllite, magnesite

Photo Credits

Most of the photos used are the author's. Numerous macro shots in the identification key were taken by Ueli Bula, www.artbula.ch, Zollikofen. Many thanks to Ueli Bula for his work.

Lukas Mauerhofer, Bremgarten, also shot a number of the macros from the author's rock collection.

The following photos are from other sources:

P. 55 upper left	Shutterstock/totajla
P. 55 upper right	Prof. Dr. Joseph Mullis, Universität Basel
P. 68 4c left	Shutterstock/www.sandatlas.org
P. 68 4c right	Shutterstock/ChWeiss
P. 69 4c bottom	Wikimedia Commons/Dentren/GNU FDL 1.2
P. 70 4e bottom left	Shutterstock/www.sandatlas.org
P. 70 4e bottom right	Wikimedia Commons/Geolina/CC-BY-SA-4.0
P. 76 7d	Wikimedia Commons/James St. John/CC-BY-SA-2.0
P. 77 7g	Martin Lind
P. 78 9a	Shutterstock/lcrms
P. 80 9d, bottom	Ian West/University of Southampton UK
P. 81 9f left	Kath. Kirchgemeinde Dübendorf/Rolf Anliker
P. 81 9f right	Wikimedia Commons/Prisantenbär/CC-BY-SA-3.0
P. 90 13d right	Wikimedia Commons/Fed/PD
P. 94 16b	Wikimedia Commons/Kevin Walsh/CC-BY-SA-2.0
P. 94 16c right	Wikimedia Commons/Hubertl/CC-BY-SA-4.0
P. 95 16e	Flickr/James St. John/CC-BY-SA-2.0
P. 100 17e left	Shutterstock/KrimKate
P. 100 17e right	Wikimedia Commons/Ra'ike/CC-BY-SA-3.0
P. 101 17i left	Shutterstock/Kicky_princess
P. 101 17i right	Shutterstock/Paulo Afonso
P. 104 18c	Shutterstock/Breck P. Kent
P. 105 18c	Shutterstock/ArtKio
P. 117 22h left	Wikimedia Commons/Hannes Grobe/CC-BY-SA-3.0
P. 122 23e right	Wikimedia Commons/Woudloper/PD
P. 125 23i, top	Shutterstock/vvoe
P. 127 24c left	Shutterstock/KrimKate
P. 127 24c right	Shutterstock/vvoe
P. 132 27c	Shutterstock/vvoe
P. 148 40b left	Raimond Spekking/CC BY-SA 4.0 (via Wikimedia Commons)
P. 148 40c left	Shutterstock/Jarous
P. 149 40d left	Shutterstock/Aleksandr Pobedimskiy
P. 149 40d right	Shutterstock/Coldmoon Photoproject
P. 149 40e left	Shutterstock/Zbynek Burival
P. 149 40f left	Wikimedia Commons/Hannes Grobe/CC-BY-SA-3.0
P. 150 40g	Shutterstock/Eduardo Estellez
P. 150 41a right	Shutterstock/Jennifer Bosvert
P. 151 41c left	Flickr/James St. John/CC-BY-SA-2.0
P. 151 41c right	MartinLind
P. 152 41e left	Shutterstock/Coldmoon Photoproject
P. 152 41e right	Shutterstock/Albert Russ
P. 154 42e left	Wikimedia Commons/James St. John/CC-BY-SA-2.0
P. 154 42e right	Shutterstock/mantisdesign
P. 161 44f	Shutterstock/KorArkaR
P. 162 44g	Shutterstock/gabriel12
P. 163 46a right	Shutterstock/isabela66
P. 164 47c	Shutterstock/futuristman
P. 165 49 right	https://gallery.yopriceville.com/
P. 167 53c left	Wikimedia Commons/Dentren/GNU FDL 1.2
P. 167 53c right	Wikimedia Commons/Ji-Elle/CC-BY-SA-3.0
P. 170 Banded Iron Formation	Wikimedia Commons/Graeme Churchard/CC-BY-SA-2.0
P. 170 Beachrock	Wikimedia Commons/Bruno Navez/CC-BY-SA-3.0
P. 170 Charnockite	Flickr/James St. John/CC-BY-SA-2.0
P. 170 Fenite	Alamy Stock Photo/Sepp Slim
P. 170 Greisen	Wikimedia Commons/Geomartin/CC-BY-SA-4.0
P. 171 Carbonatite	Wikimedia Commons/Eurico Zimbres/CC-BY-SA-3.0
P. 171 Kimberlite	Wikimedia Commons/Woudloper/CC-BY-SA-1.0
P. 171 Komatiite	Alamy Stock Photo/Adwo
P. 171 Phosphorite	Wikimedia Commons/Ra'ike/CC-BY-SA-3.0
P. 171 Suevite	Wikimedia Commons/Martin Schmieder/CC-BY-SA-0 1.0Index

Index